RAINBOW DUST

Rainbow Dust

Three Centuries of Delight in British Butterflies

PETER MARREN

SQUARE PEG

1 3 5 7 9 10 8 6 4 2

Square Peg, an imprint of Vintage,
20 Vauxhall Bridge Road,
London SW1V 2SA

Square Peg is part of the Penguin Random House group of companies whose
addresses can be found at global.penguinrandomhouse.com.

Penguin
Random House
UK

First published by Square Peg in 2015

www.vintage-books.co.uk

A CIP catalogue record for this book is available
from the British Library

ISBN 9780224098656

Text designed by Lindsay Nash

Typeset in Fournier MT by Palimpsest Book Production Limited

For Emma and Claire Garnett

If a child is to keep her inborn sense of wonder,
she needs the companionship of at least one adult
who can share it, rediscovering with her the joy,
excitement and mystery of the world we live in.

RACHEL CARSON, *The Sense of Wonder*

A Note on the Illustrations

The images of butterflies at the head of each chapter were taken from *Papilionum Britanniae Icones* by James Petiver (1717) and *An Illustrated Natural History of British Butterflies* by Edward Newman (1871).

Contents

INTRODUCTION

The Painted Lady

All one's butterfly memories are sunny ones,
bright pictures in the mind which colour
the dark days of winter.

BB (DENYS WATKINS-PITCHFORD),
Ramblings of a Sportsman-Naturalist (1979)

My friend Rosemary was a farmer's daughter. She was six and
I was five and so we were old enough to be let loose in the big
walled garden. We zoomed about on our trikes, took turns to
push each other on the big swing-seat and played dodge-and-
dare with the lawn sprinkler. It was hot. I faintly remember the
heat shimmer and the baking orange bricks of the wall and the
fragrance of honeysuckle. But one moment, which was over
and done with in less than a minute, remains stuck in my head
like a freeze-frame. I caught a butterfly. That was all. A brown-
and-white butterfly. I remember it, I think, partly because it
reminded me that I was short-sighted. At that time I hadn't
told anyone. It was a secret I intended to keep for as long as
possible and in the meantime I had grown used to seeing the

world in soft focus. Fellow myopics will know what it is like when something suddenly swims into view inches from your face. It comes up at you, first in the usual blur and then quite suddenly snaps into focus revealing its form, its solidity, its hard-edged outline. Over perhaps a second this butterfly changed from a formless motion to a soft, fawn-coloured blur. And then, as it settled and flattened its wings against the hot brickwork, it turned into a Painted Lady. It had flesh-brown wings flushed with salmon-pink in a pretty latticework of black, like a glimpse of a sunset through a thorn bush. The black tips of its forewings contained a circlet of white spots, like dabs of snow melting on a slate. I knew right away what it was. I was keen on nature even at that age. I picked up shells and bladderwrack from the cold Yorkshire beach, and on the white-painted ledge of the kitchen window was a jar of tadpoles which I hoped would one day turn into frogs. But I had never taken much notice of butterflies before, mainly because I couldn't see them. I liked this one, and, like any 5-year-old, my first impulse was to try and catch it.

Young children learn things without conscious effort. They are hard-wired genetically to learn fast when little. I thirsted for knowledge, not in order to survive in a dangerous world like our cavemen ancestors, but for its own sake. I was starting to find the natural world a thrilling place, full of wonders and shocks. In the 1950s the commercial world came to the aid of boys like me by offering amazing facts on the backs of match-boxes and cereal boxes, or on the wildlife cards given away with every packet of tea. Even comics often managed to sneak something about animals or the planets in between 'Tough of the Track' and 'Roy of the Rovers'. Maybe my unconscious absorption of the names of butterflies came from there or perhaps from one of those big educational books you got for your birthday with titles like *The Wonderland of Knowledge* or *The Adventure Book for Boys*.

The hands of 5-year-olds are damp and clumsy. When the butterfly slithered and shot out from between my fingers, I noticed that some of its colours had been left behind on my sweaty palm, like the imprint of a chalk drawing. I held my hand up to the sun and watched the particles shine and flicker: orange, yellow and black, in shimmery, summery specks. 'Look!' I yelled. Rosemary tricycled over and looked. 'It's like transfers!' There was a craze just then for transferable film which came in penny sheets and which you could stick on to your arm like tattoos. 'No', she disagreed. 'It's smudgier than transfers. It's butterfly dust. Angel dust!' I rubbed the tiny scales together with my finger, mixing up the colours into a mud-like smear. A few sparks still glinted from my fingertip. 'Rainbow dust,' I said at last. 'That's what it is. Rainbow dust.'

It was, I realise now, a Nabokov Moment. Of all the inconsequential things that happened to me when I was 5 – falling down the rockery, piling cold Yorkshire sand into my Mickey Mouse bucket, snivelling over my uneatable school dinner – this one stays bright after all the others have faded. Vladimir Nabokov was the great Russian-born novelist, author of *Lolita*, who also had a passion for butterflies. It was a Nabokov Moment because only he could put into words what most of us can only feel: the frankly sensual moment in a child's life when the full force of nature is felt for the first time. It marks the first time when, to your vast surprise, the colours and scents and sounds of the natural world seem to imprint themselves inside you, not only in your mind but in the very marrow of your bones. Nabokov called it ecstasy, but behind the ecstasy there is something else, something nameless and hard to explain: 'It is a momentary vacuum,' suggested Nabokov, 'into which rushes all that I love. A sense of one-ness with sun and stone.' When something like that happens, you feel it has been given to you. Nabokov described the feeling as 'a thrill of gratitude to whom it may concern – to the contrapuntal genius of human fate or to tender ghosts humouring a lucky mortal'. Call it what you will, the Painted Lady that bore my thumbprints on its wings represented something significant in my young life. It represented the point at which my life as a naturalist began.

Wildlife for short-sighted, untruthful little Peter needed to be the sort I could hold in my hand, or in a jam jar; shells cast up on the muddy shore, a posy of flowers plucked from the edge of the cornfield, those imprisoned taddies that never did make it into frogs. Flowers and shells did not fly away, unlike the birds which chirped and fluttered in the fog beyond my damaged vision. But I could see insects, just so long as they stayed still for a moment when I loomed near. After a while I caught the knack of stalking butterflies: of moving softly, taking care of my shadow and never for a moment taking my eyes off those velvet wings with their own sightless 'eyes'. There is, I realised, a quiet thrill in knowing you are invisible.

When I grew older I started to collect and rear butterflies, first in Britain and later in Europe. In those days, collecting butterflies was just another of those things one did as a boy, either alone, or with my brother and my dad, or with some of the lads from the village. It was on a par with pond-dipping with net and jam jar, or climbing trees to inspect bird's nests.

Eventually I got bored with my other hobbies: my chemistry set, my model railway, the Airfix ships and planes, those half-built Meccano cranes. But butterflies must have impressed me at a deeper level for they alone accompanied me into my teens. At length I outgrew collecting for collecting's sake but I never lost an unnameable feeling for the living world. It drew from me wonder, excitement and intense curiosity as well as simple happiness at watching other forms of life. Collecting butterflies also produced more negative emotions: greed, for instance, and shame for having killed such lovely, innocent things. By then I also envied butterflies their freedom and ability to indulge their own pleasures while I was stuck behind the gates at my boarding school. I remembered then the butterfly's words from one of Hans Christian Andersen's fairy tales: 'Just living is not enough. One must have sunshine, freedom and a little flower.'

Other insects do not seem to have this emotional impact. For those bitten with the bug, butterflies feed the emotions in strangely intense ways, perhaps in similar ways to paintings or music or a good book. I suspect many people will recognise what I mean yet share my difficulty in expressing it in words. You respond to something with real passion but you don't really know why. It becomes a part of you and part of the way you spend your time and live your life. It was through butterflies, not school lessons or even books, that I discovered natural history. It came about from watching them with my own pair of malformed, short-sighted eyes.

When I was a boy there were relatively few butterfly books on the market – and most of those were re-treads of books written decades earlier. Today there are dozens, from baldly factual field guides to Patrick Barkham's enthralling quest to see all the native butterflies, *The Butterfly Isles*. Yet most of the literature about butterflies is scientific: it is about facts, whether it is their life cycles or their behaviour or how butterflies survive long enough to pass on their genes to the next generation: the sole reason for every butterfly's existence.

I do not intend to repeat those facts but to look at butterflies from a different perspective: the effect of butterflies on ourselves and the way they affect our thoughts and inspire ideas; in other words, and for lack of a better phrase, their cultural impact. For more than 300 years British butterflies have been hoarded, bred in tins and jam jars, painted, thought about, written about, wondered at, recorded and, more recently, conserved, often with rare dedication, skill and passion. Where does such devotion come from? Even a boy's simple butterfly collection fits into this widely shared feeling that butterflies are special – for much trouble and artifice went into those bedraggled rows of wonkily pinned specimens. What is it about butterflies that have the power to move us, to give them wonderfully poetic names both in English and in Latin, and to keep them by us in books, on birthday cards or in printed or painted images on the wall? What is their secret?

This book is my attempt to write a personal 'cultural life' of British butterflies and to try to give a sense of their lasting appeal. It begins with my own faltering effort to build a butterfly collection and ends with the communal efforts of conservationists to save them in a world that has turned hostile to wildlife. Sandwiched in between is a journey through the ways in which butterflies have made us think: about images and art, about superstition and advertising; and about the way we relate to the countryside and the natural world. My originally chosen subtitle 'Where people met butterflies' sums up this interaction between the butterfly and its observer: a relationship and a shared history. Along the way are portraits of people I have known or admired or whose lives, like mine, were changed by encounters with butterflies.

My title, *Rainbow Dust*, based on that distant encounter with a Painted Lady in a Yorkshire garden, provides a hint that butterflies are not all about hope and joy and human aspiration. As George Monbiot expressed it recently, 'to love the natural world is to suffer a series of griefs, each compounding the last'.[1] Butterflies, along with bees, moths, mayflies and many more wonderful insects, are in decline. Moreover, probably no boy will ever again shoulder his net and bag and ride off into butterfly country with the same sense of freedom that I, and many generations before me, were lucky enough to have had: 'never glad confident morning again'.[2] Our reinvented Eden has 'don't touch' signs. Increasingly, encounters with butterflies do not just happen but must be planned for, perhaps involving a car journey followed by a purposeful walk. But we can still let butterflies into our minds and our thoughts. Just as they were in my childhood, and long before that, butterflies represent a doorway: the key to seeing the world beyond through the eyes of a naturalist.

1.

Meeting the Butterfly

Everybody knows what a butterfly is. They are the most colourful of insects, the ones with painting in their wings. The butterfly flies in the sunshine and visits flowers and, at least in the eyes of the poet, lives a short but joyous and carefree life. They charm and delight us and make us think well of nature. It is their beautiful colours that might have inspired the name 'butter-fly' and that certainly underlie their individual names. In Britain we have Common Blues and Green-veined Whites, Clouded Yellows and Wall Browns. Those with many colours or more complicated patterns we accord names like tortoise-shells, peacocks and fritillaries. We award high rank to some: admirals, monarchs, emperors.

Butterflies are a self-contained group of insects, a distinctive class within the order Lepidoptera ('scaly-wings') which also contains the much more numerous moths. It ought to be easy to say what a butterfly is and how it differs from a moth. But a scientific definition of a butterfly is less clear-cut. Yes, they fly by day, but so do a great many moths. Yes, they are colourful but have you ever seen a tiger-moth? Yes, most butterflies do not resemble your average moth but the small butterflies called skippers are very moth-like. Some moths have stout, hairy bodies but others are more slender and butterfly-like.

The one sure way to tell a butterfly from a moth is to look at their antennae. These are used to detect scents in the air, such as the delicate pheromone of a mate or the chemical signature of a favourite flower. All butterflies have little clubs at the end of their antennae. Moth antennae, on the other hand, can be feathered, or like wires, but are never clubbed – though, just to be awkward, the day-flying burnet moths have antennae that thicken towards the ends. There are other technical differences such as the way the wings are constructed. Most moths possess a little hook and attachment that links the wings and enables them to buzz like bees. Butterflies do not have that option; instead, and, again, with the usual exception of the skippers, they beat their wings or, in some cases, glide between wingbeats. Butterflies 'flutter by'. Moths buzz off.

The simplest way to tell a butterfly from a moth is to get to know the butterflies. There are only fifty-nine resident species in Britain, plus eight rare migrants and three extinct species, compared with nearly a thousand larger moths alone (and still more tiny ones). Butterflies that visit gardens are even fewer – perhaps twenty. Nearly all of our species are easy to recognise and they all have English names. And when it comes to identification we are spoiled for choice. It would be hard to find any bookseller without at least a couple of butterfly books. Even guides to insects in general routinely include all, or nearly all, the butterflies. There are also numerous sites online, as well as gizmos such as CD-Roms that picture them all and point out the differences.

Every species of butterfly has its own lifestyle. Some live mainly in woods, others in grassland or moorland or along the coast. Some, generally those with a powerful flight, roam from place to place while others tend to stay put, close to the bush or patch of flowers where they came into the world. What they all share is a life cycle that involves a series of transformations. The female butterfly lays an egg which hatches into a tiny caterpillar. A caterpillar has only one biological task: it must eat and carry on eating until it reaches full size, and as rapidly as possible. Caterpillars are plant eaters, though the exact plant they eat differs from species to species. Peacocks and Red Admirals feed on stinging nettles. Many fritillary caterpillars graze on violet leaves. Most of the browns and skippers feed on wild grasses. Being of only limited flexibility, caterpillars have to pause and change their skin every now and then as they fatten up. In the final skin change, they change form completely and turn into the dormant phase known as a pupa or chrysalis. Inside the apparently sleeping chrysalis a rearrangement takes place that eventually produces the adult butterfly. A day or two before the butterfly is fully formed, the chrysalis darkens and wing colours show faintly beneath the skin. Finally, when all is ready, the chrysalis splits open and, in one of those moments in nature that we find so miraculous, the butterfly pops its head out and sees and smells the world for the first time.

Anyone who has reared butterflies will know that moment. Caterpillars have limited value as pets. They show no emotions at all, let alone evident happiness or gratitude as you supply them, over and over, with fresh leaves. They simply chew plant tissue at one end and void it at the other in little brown pellets known as frass. Wherever possible I used to pot some plants or cover a branch with muslin, drop in the caterpillars and forget about them. Admittedly not all caterpillars are dull; many are adorned with pretty spikes or horns, or attractive colours, and the Puss Moth caterpillar even has a pair of little tails. And there is something hypnotic about watching a big caterpillar – a hawkmoth, say – quickly reduce a leaf to a bare stalk. But on the whole it is a relief, not to say an end to anxiety, when it finally reaches full size, spins a pad of silk, hunches and seems to shrink into itself and duly turns into a pupa or chrysalis.

The emergence of the butterfly is something else. I see it in my mind's eye now, that moment when the sleeping chrysalis seems to wake up, wriggle and kick, and suddenly split along the groove where the wings show through. The thing inside, not yet recognisable as a butterfly, puts out its head. It unfurls two long antennae from a pouch just below the skin of the chrysalis and, flickering like a snake, tries out its coiled tongue. Quickly, almost too quickly to take it all in, the thing wriggles out, using its new-born feet to grip the old skin and drag the rest of its body out of its shell. Attached to its chest, just above the legs, are four pulpy, coloured flaps. The thing reverses itself and, now facing up instead of down, and gripping the shell of the chrysalis, it sets about creating and drying its butterfly wings. Its body bulges and squeezes as fluid is pumped down the veins to the edges of the wings. Ever so slowly they unfurl and expand. It might take an hour and in every minute the thing becomes more and more recognisable. Then, in a moment of glorious apotheosis it opens its pristine, gorgeous wings for the first time and turns into a butterfly. Any excess fluid, which can be transparent but in some species is blood-red, is voided. And then, after a couple of practice beats of its splendid wings, the butterfly leaves its old body behind and launches out into the air and begins its brief adult life.

How brief is that? Butterflies live longer than mayflies but not as long as bees. For those that do not perish prematurely, in a spider's web or as bird food, their adult life may last a week or two. The colours and patterns of a butterfly's wing are formed by tiny overlapping scales. The silky sheen of a newly emerged butterfly is lost within a day or two, and, as more and more of the scales are rubbed or shed it becomes increasingly worn and faded. The wings also become ripped and tattered from rain and thorns or take on triangular tears from bird attacks. The longest-lived butterflies are those that pass the winter as adults, in hibernation. Such are the Small Tortoiseshell and the Peacock, which often seek refuge in garden sheds, and the Brimstone which prefers to overwinter in a dense knot of ivy. Those species can live several months, though, of course, they are inactive and 'asleep' half the time. If you add up the time they spend as an egg, caterpillar and chrysalis, single-brooded butterflies, such as the Purple Emperor or the Chalkhill Blue, live a full year. Those which have more than one generation, like the Adonis Blue, which appears in May and again in August, have shorter cycles. A full summer cycle from egg to egg-laying butterfly takes about three months.

How does a butterfly spend its life? I once helped to supervise the work of a student who did her best to follow one species, the smallest of all our butterflies – the Small Blue – from the day it emerged from the chrysalis until the day it died. A large part of this butterfly's life was spent immobile on a grass stalk ('sleeping'), warming up in the sun (sunbathing) or gorging on the nectar of vetch flowers. Butterflies taste with their feet. It might sound odd but this gives them a much more sensitive palate than the human tongue. Like many butterflies, this feisty little blue established a home range or territory. It saw off rivals which it detected by sight and found a mate which it discovered by using its sensitive antennae. The mating ritual of a butterfly is unhurried. The male displays its wings, might waft an attractive scent towards its partner with its wings and generally does its best to impress. The female might offer herself or she might not. If the former the couple stay paired for an hour or more. There is no hurry.

It all sounded like a pretty good life to me (though with the danger of sudden death at any moment) and I felt almost envious of the Small Blue. The female of the species, which in butter-flies is often less brightly coloured, has the tougher role. She has the burden of scores of thick-shelled eggs to carry and the chore of finding places to lay them. In the case of the Small Blue, the butterfly uses its senses to find a suitable patch of kidney vetch – its yellow blossoms are the small Blue caterpil-lar's only 'food plant' – that is, one not already discovered by a rival Small Blue. The competition must be intense. Each time she has to deposit a single blue-white egg in exactly the right place, on the hairy calyx leaf underneath the flower. This involves abdominal straining as the butterfly does its best to glue the egg into a hiding place while clinging on to the awkwardly shaped flower. At the end of her mission she must be exhausted. On the whole I would rather be a male Small Blue – fucks, fights and scented delights! – even though they have an unfortunate appetite for dog shit.

Butterflies feed on liquids using a long tubular proboscis, a kind of flexible straw. Their standard diet is sugar-rich nectar from flowers – and they incidentally pollinate the flowers in the process. Some species are also attracted to 'honeydew' left by aphids, to salt-rich water oozing from seepages in the bare earth, the juices from overripe fruit, sap escaping from a wounded tree or fresh mammal droppings. In very hot weather butterflies can even be attracted to sweat. Needless to say, they need a plentiful supply of the right kind of flowers. The one way humankind can help them is to provide suitable flowers in our gardens, such as mint, lavender, wallflowers, red valerian and, of course, the 'butterfly bush', buddleia. It is easy to design a butterfly garden: just plant the right species in a sunny location. There is no need to plant stinging nettles. There are plenty of them around already.

Butterflies face many hazards: parasites, disease, predators, plus our own contribution of insecticides and habitat destruction. The chances of any one egg making it through to an adult butterfly are small. That is why butterflies lay a lot of eggs (by the same token, stealing a butterfly's egg does not have the same impact as stealing a bird's egg). Britain offers a further problem in that we have a cool, wet, uncertain climate. Butterflies thrive on sunshine and certainty. Unlike mammals, they have no internal heating system. Their body heat is generated externally by basking and using their wings to regulate the temperature. A spell of cold weather results in more caterpillars becoming casualties and so fewer butterflies on the wing, and therefore fewer eggs being laid. Bad summers are very bad for butterflies. Fortunately they have the capacity to recover quickly in warm years (otherwise we would probably have no butterflies at all).

Like birds, some butterflies migrate (though, unlike birds,

that fact was not appreciated properly until well into the twentieth century). Among our common migrants, the ones that arrive every year, and sometimes in large numbers, are the Red Admiral, Painted Lady and Clouded Yellow. We also have several more occasional migrants such as the Camberwell Beauty and the Monarch. To fly long distances over land and sea takes strong wings and tough flight muscles. You can see it at work when you encounter a Painted Lady on its way north. These butterflies do not flutter by; they fly straight and fast, as though they know where they are going (each has, in fact, no idea where it is heading, but its genes do). Butterflies are short-lived and few, if any, of those that reach us in the spring will survive to attempt the return journey to southern Europe or North Africa. It is the second, British-born generation that makes the return journey. You can see migrant butterflies as conveyors of genetic material, taking their genes north for the summer and then, in a different body and generation, back home again. But even when the facts are familiar and the process broadly understood, it still seems incredible that anything as fragile as a paper-winged butterfly can endure weeks of flight, using the sun as a compass, high above the curvature of the earth.

Migration is not the only butterfly mystery still being unravelled. Another is how dependent many of them are on other species, particularly ants. Ants are aggressive predators which live in underground nests in very large numbers. Make friends with an ant and you have a protector for life. The dependence of one of our rarest butterflies, the Large Blue, on a particular kind of ant has been well known for a century. Here the protection is rather lopsided since the butterfly's caterpillar helps itself to the ant's grubs as a midwinter larder (helped, it was found later, by smelling like an ant and even 'singing' like one). But other blues and hairstreaks, too, depend on the protection of ants. Their caterpillars and chrysalids produce secretions that attract ants and ensure they become well-attended by ant escorts. One of the most amazing sights of the butterfly world, one discovered only recently, is the emergence from the ant's nest of a Silver-studded Blue butterfly with anything up to eight ants frantically licking its body. The hour which every butterfly has to spend drying its wings is perhaps the most vulnerable of its whole life. But with a posse of fierce ants to stand guard, the chances of survival are much greater.

Insights like this take many years of patient research and observation. Butterflies are amazing. Nature is amazing. But anyone can get a front seat on the drama of butterfly lives, simply by rearing them. I used to do that. I used to collect them too. I'll explain how and why in the next chapter.

2.

Chasing the Clouded Yellow

'I haven't seen a single butterfly collector in the past twenty-five years', said Matthew Oates. We were staring up at the treetops hoping to spot a Purple Emperor. Purple Emperor butterflies spend most of their lives up in the sunlit canopy of English woods, descending only to drink from puddles or to imbibe some life-enhancing substance from dog shit or roadkill. Unless you are in luck, you need binoculars to get a good view of the dark butterflies forty feet overhead. Emperor-watching is like birdwatching: a matter of glimpses punctuated by the occasional freeze-frame as the glorious purple-shot butterfly alights on a leaf to bask or sip the sticky honeydew secreted by incontinent aphids.

Matthew is our most experienced observer of the Purple Emperor, once the greatest prize of any collector. And if Matthew hasn't seen a single person trying to catch his beloved Emperors in a full quarter-century then they are either very good at hiding or, more likely, no one is trying to catch them any more.

Like Matthew, I cannot remember when I last witnessed

anyone collecting British butterflies. But I am old enough to remember when it was still one of the classic hobbies of a young naturalist, along with other kinds of collecting – fossils, pebbles, feathers, pressed flowers, shells. Those hobbyists included me; I too was for a while addicted to catching butterflies. Back then, in the 1950s and early 1960s, almost every butterfly book offered tips on how to form your own collection and how to breed up more butterflies from eggs and caterpillars. In fact it was hard to imagine anyone being seriously interested in butterflies without collecting them.

It was maybe ten years after that when attitudes began to change and collecting was no longer respectable. It wasn't through any change in the law. A few species of butterflies and moths *were* legally protected for the first time in the 1970s but the intention behind the protection was educational. It drew attention to the plight of certain rare and attractive species that could, in theory, become endangered by over-collecting – but offered no evidence that anyone was actually collecting them. Perhaps the demise of the butterfly collector had more to do with improved camera technology and faster colour film which made it much easier than before to obtain good, sharp images of live butterflies. It was also, of course, connected with a dawning realisation that we were trashing the countryside and its wildlife. Butterflies were becoming less common, not because of collecting but as a result of their habitats being ploughed or drained away or turned into roads and suburbs. The same kinds of people who collected butterflies two or three generations ago became equally ardent conservationists. They stopped collecting butterflies and joined a recording scheme instead, or acted as volunteer wardens on their local nature reserve. Conservation was in and collecting was out. Quite suddenly, and without much fuss or argument, it had become socially unacceptable, and that was that.

So I belonged, in all probability, to the last generation that could set out on foot, or on a bike, with a net and satchel, without feeling self-conscious. We were the last of a line that stretches back at least 300 years to the dawn of the Enlightenment. Butterfly collecting is often thought of as one of the eccentric things the Victorians did – a case of butterflies being a standard prop in BBC adaptations of Trollope or Dickens. But the hobby was already old by then. It was probably John Ray, the first of the English parson-naturalists, who, along with his family and friends, was the first to collect butterflies and their caterpillars in a purposeful way, in the lost England of the 1690s.

For nearly three hundred years, collecting and rearing went hand in glove with the study of insects, known, since the early nineteenth century, as entomology. It was hard to make sense of insects, or paint their portraits, without a reference collection – and for some of the more challenging groups of insects that still holds true. As for any possible moral issues involved, Moses Harris, author of the most popular butterfly book of the eighteenth century, *The Aurelian* ('aurelian' is another word for butterfly collector) had the answer. He found it in the Bible in the words of Psalm 111: *The works of the Lord are great, sought out of all them that have pleasure therein.* The reason why God sent us butterflies was obvious: it was for the joy they give us. And as Moses Harris hinted in his celebrated frontispiece, showing an improbably well-dressed collector at work with his big, baggy net, taking an interest in butterflies was akin to seeking to know the mind of the Creator. That, since

Isaac Newton, had been a respectable and praiseworthy aim. And besides, a cased arrangement of butterflies looked well on the study wall.

Collecting butterflies was, and is, a complicated business. You needed a lot of kit: a net, round boxes, papers and boards to set the wings in position, and cork-lined, bug-proof boxes to store the collection. You needed custom-made pins to mount the specimens neatly and a pair of special tweezers ('entomologist's forceps') to handle them, as well as a sponge-lined tin to 'relax' the wings of dried butterflies. Once you had passed the jam-jar stage, you wanted at least a couple of muslin-lined wooden cages to rear the caterpillars, or, if that was beyond your price range, a home-made substitute using a cake tin and a cylinder of polythene. And of course you needed something sufficiently deadly to kill a butterfly quickly and humanely without damaging its delicate wings.

By my time chemists had stopped supplying schoolboys with cyanide but we used something almost equally unpleasant – carbon tetrachloride, widely used then as a degreasing agent for home dry-cleaning but now known to be a dangerous carcinogen. An accidental whiff of 'carbon tet' made your lungs go cold. Some boys used household ammonia, which brought tears to your eyes every time you opened the bottle. Later on we used ethyl acetate, which smelt a bit like pear drops but which has been banned by the EU all the same.

Spreading the butterfly's delicate wings without ripping them or leaving your thumbprint behind in a smear of scales took practice. It has been compared with unfolding an origami figure. Even with a boy's nimble fingers, well practised on making plastic models of Spitfires and battleships, butterflies were a challenge. God knows how many we wasted before there was a single one we could look at without shame. The job had to be neatly done with the pin stuck exactly through the centre of the thorax at ninety degrees to the spread wings. After leaving the specimen to dry on the setting board for a few weeks, you removed the now brittle corpse with great care and added a label, stating the place and date of capture. This turned it from a mere object into a scientific *specimen* with its own *data*. If you felt the job was good enough to acknowledge, you added your name to the tiny square of paper: 'On buddleia. Longstanton. 10.8.63. Peter Marren.' For butterflies you had bred yourself, you added the mysterious letter 'x': 'X. On nettles. Longstanton. Marren.' That meant it had been reared from an egg or a caterpillar and not caught in the wild. Finally, and with some small sense of pride, you pinned the labelled butterfly into your storebox, a hinged wooden box lined with cork and white paper and fumigated with naphthalene – which, we know now, but didn't then, is yet another dangerous carcinogen. We collectors lived among toxic fumes.

The smell of naphthalene, otherwise known, with a certain irony, as mothballs, still brings back memories of those distant days, just as a tea-soaked madeleine cake ignited that 'vast structure of recollection' for Marcel Proust. It's the smell of a lost world, of home museums, of a time when one's excitement at discovering the natural world was just beginning. Everything about nature was fresh and wonderful, and a well-stocked naturalist's den was the most worthwhile thing in the world.

Perhaps you think that I must have been a strange sort of boy to want to do this when I could have been out there with my mates kicking a ball about. But a surprising number of other boys in my village shared the passion, at least for a while. Collecting back then seemed as natural as birding with binoculars today. There was nothing to stop you except on nature reserves – and there weren't many of those. Even on television, nature programmes were mostly about zoos, arguably another form of collecting, or treks to the far side of the world to catch more animals for the zoo. The best-known butterfly man on television and radio was L. Hugh Newman who made a living breeding butterflies and moths for collectors. None of us thought we were doing any harm; quite the contrary. But the focus was on insects as a hobby; it was for pleasure and personal fulfilment, and we had never heard of ecosystems or what would become known as 'the environment'. The past, even the recent past, is a faraway country. We did things differently then.

Collecting led straight to entomology, the scientific study of insects, and, more broadly, into what was then called natural history. There was an adventure side to it too – chasing the golden butterfly took you out into wonderful, wild places you would otherwise never have found. The hobby was surprisingly gregarious and it put you in touch with like-minded people. It was fun to meet fellow enthusiasts and gossip about your shared passion. Such chance meetings often ended with an exchange – a jar full of caterpillars, say, or a couple of hawkmoth pupae swathed in cotton wool. I was a member of the youth group of the Amateur Entomologist's Society that met at Holland Park School in London once a year. The passion taught you to recognise the plants on which the caterpillars fed, and also to notice that other insects, such as hoverflies, visited the same flowers. You learned how to look and how to see. It becomes an ingrained habit after a while. That's one way you can spot a naturalist. They are always looking.

I knew the names of the British butterflies long before I started collecting them. Somehow my unconscious mind seems to have absorbed them without any real effort of memory, just as other boys know the names of all the players, past and present, in the football team they support. I enjoyed the Red Admirals and Peacocks visiting the Michaelmas daisies and the whites fluttering over Dad's Brussels sprouts. I had heard about the wonderful Purple Emperor who rules his insect kingdom from his oak-tree throne, and I dreamt about the elegant black-and-yellow Swallowtail, soaring and dipping above the Norfolk reeds. Yet when the collecting bug first hit me, my thirst was not for butterflies, oddly enough, but for their dusky and far more numerous cousins, the moths. It came about the day I bought *The Observer's Book of Larger Moths* by R. L. E. Ford, a pretty useful book you could buy with pocket money – it cost five bob, or 25p.[1] Unlike butterflies, I knew nothing about moths then and remember being amazed at what a good-looking lot they were. There were noble hawkmoths shaped like torpedoes with the wings of fighter planes, the Oak Eggar, furry and brown as a winged teddy bear, and underwing moths with unexpected bright colours on their hindwings, red, yellow and crimson, like the flash of naughty knickers beneath a raincoat. A light came on behind my greedy little eyes but, since it was winter, the great moth hunt I anticipated would have to wait. But I knew there and then that this was going to be something big.

My idea of moths was formed by those pictures which showed them as dead museum specimens with their wings spread out in an unnatural way. One reason many find it hard to identify them from such pictures is that living moths look nothing like that. In most cases the wings are folded neatly over the body like a well-worn overcoat. The moths in my first collection looked roughly halfway between museum specimens and living insects. Their wings were all over the place: a horrible row of drooping specimens in various defeated attitudes. Moths proved harder to track down than I had thought and the dead ones I pulled from cobwebs or found floating in the toilet bowl were nothing like the lovely things in the *Observer's Book*. I've seen more attractive roadkill. Then I got some lads to join me in a great village moth hunt. We searched the porch lights and rummaged in garden sheds and outhouses and, in the end, and more or less by accident, we finally ran down a fairly decent-looking moth. It was a Wood Leopard, looking just like the one on the last page of the *Observer's Book*, with white wings spotted with black and a spiked tip at the end of its body indicating that this moth was a female (the spike is an ovipositor, used to deposit its eggs inside cracks in the bark). It was newly emerged and drying its wings on the trunk of a plum tree. I can see it now, long and sleek with furry legs and the still slightly damp wings folded over its plump abdomen. We looked at it for a bit wondering what to do. I had not thought very much about how to kill a moth without damaging its beauty. In the end, and it shames me to say it, a tough boy called Lefty decided that the occasion called for his sheath knife. By the time Lefty had finished carving up the poor Wood Leopard none of us wanted it any more. That might well have put an end to my craze for moths – but for one thing.

Somehow I discovered that the same Mr Ford who wrote

my *Observer's Book* (but had neglected to tell me what to do with a Wood Leopard) ran a natural-history supply company called Watkins & Doncaster. I got hold of their catalogue and my eyes nearly popped out of my head. I thought it was possibly the most wonderful thing I had ever seen. Everything a naturalist could want was in there along with other items I hadn't realised even existed. There were special devices for trapping plankton and a folding tray for catching caterpillars and a great variety of nets from canvas 'sweeping nets' to huge professional-looking kite nets, so-called because the top was shaped like an old-fashioned kite. There was even an instrument for scooping the brains out of any small, dead mammal one wanted to stuff. I knew now that collecting moths required a hurricane lamp and a bottle of Watkins & Doncaster's patent treacle mix, apparently an infallible draw for moths when painted on to tree trunks or park benches.

Meanwhile, I noted with interest, you could kick off your collection by buying moths that had been set by someone rather more skilled than oneself. Watkins & Doncaster sold pretty well all the moths in the *Observer's Book* ranging from a little one called the Chinese Character costing 2d (rather less than a modern penny) to the huge Death's-head Hawkmoth at an unaffordable 12/6 (about 60p). God, I so wanted a Death's-head Hawkmoth. Since this was going to mean a serious outlay of pocket money, I resorted to moral blackmail. I begged my father to let me have my birthday in advance, so to speak, oh and Christmas too, preferably. Dad, I reasoned, owed me a big favour. I was about to be sent away to boarding school and I knew I could count on his feelings of guilt. I reminded him that I would not be at home for my real birthday, would I, *Dad*.

So my poor parent gave in, and, one happy weekend, we set off in the family Morris, me and my mate Jonathan, to find Watkins & Doncaster. This took some doing because all we had to go on was the sketchy map on the back of the catalogue. But, after a few unscheduled turns around Shooter's Hill, we spotted it, an unprepossessing house halfway down a long avenue. Here was the Moth Shop at last. I leapt out and unhesitatingly knocked on the door of the author of *The Observer's Book of Larger Moths*.

All things considered Mr Richard Ford was very patient with his unexpected visitors. He invited us in and it was indeed a place of wonders, with butterfly nets stacked like umbrellas, cases of tropical butterflies and silk moths on the walls, and more butterflies and moths in a row of antique-looking cabinets lined against the inside wall in a gallery running along the back.

In better times the premises had been in the heart of London. The firm's logo of a Swallowtail had hung proudly on a sign from a tall Victorian tenement on the Strand, high above the roaring traffic. To reach its cramped quarters you had first to squeeze past a barber's shop and then clump up a dusty wooden staircase. Old Mr Doncaster, who had run the business for nearly half a century, could neither speak nor hear. His customers wrote down what they wanted on a slate slung round his ancient neck.

Richard Ford had bought the business in 1939 and turned it in a more modern direction, away from eggs and stuffed birds and towards entomology and field study. Then came the fateful day when the old building was condemned to be pulled down as part of a traffic-widening scheme for the Strand. Now exiled in Welling, in the east London suburbs, Ford must have missed the naturalists who would drop in for a chat and an exchange of entomological gossip. Watkins & Doncaster was now, necessarily, mostly a postal business. We might have been his first visitors for quite a while, but Jonathan and I were in a hurry to see the moths and in no mood for social niceties. We were ushered into the gallery to inspect the naphtha-scented cabinets. We drooled over the mothy delights inside, pinned in row upon row, most of them set by collectors from earlier in the century. We were both on a budget and had already worked out which ones we wanted most. Richard Ford pulled open one drawer after another while we said things like, 'Um, maybe I'll have that one, no, oh no – you can put that one back, Mr Ford – I'll have that one, no, no, not that one, *that* one,' and so on, until my box was full of pinned and labelled specimens, variously banded, tufted and spotted. But Ford was a naturalist, not a mere tradesman. 'Any fool can *buy* a moth,' he pointed out, rather sniffily. What you should be doing, he continued, keen young lads like you, is searching for caterpillars, digging around trees for pupae, smearing the trees with treacle, *getting to know about moths*. Yes, yes, I thought, but how much searching and digging and painting with treacle would I have to do to find a fine Spurge Hawkmoth like this one? More than three bob's worth, I bet.

Without an expensive light trap it clearly was going to be a lot of trouble to catch moths. Butterflies, on the other hand, were more accessible, and you could hunt them by day, with your mates. My younger brother Chris, who could never see what I saw in moths in the first place, was already keen to start. And so it was to butterflies, after all, that I turned when spring came round again. I still have the first butterfly I managed to set neatly: a female Peacock with sloping wings and a smudge on one of them, and the antennae looping forward like antlers. I practised my setting on Peacocks and tortoiseshells which are relatively easy to get right. It was a while before I could master the more delicate Orange-tip, and quite a while before I could set a blue or a skipper without leaving rips and tears in the wings through which the light would shine reproachfully. But practice makes perfect and I got there in the end.

Meanwhile a handyman who worked for Dad managed to knock me up a cane net for which Mum sowed a bag made from net-curtain material. I stowed my collecting stuff in a canvas Second World War-vintage gas-mask bag slung on a strap which might have been made for the purpose. Various village lads produced their own eccentric versions of a butterfly hunter's kit with similarly home-made nets stuck on walking sticks or broom handles. Thus equipped we roamed the village hinterland in a pack. Woe betide any decent-looking butterfly that came anywhere near us. For a few bright weeks we were in butterfly heaven, starting with the Orange-tips fluttering over frothy lanes of cow parsley, then moving on to the tortoiseshells in their puddles of sunlight on the garden path and the odd Holly Blue, jinking through the shrubbery in a twinkle of blue and white.

In the way of things the other kids lost interest one by one. All too soon my family had to move from that bright butterfly village to a barren new posting in north London. I spent the term-time locked up in boarding school where I could only dream of butterflies. Someone whose father had collected butterflies in Malaya gave me a small collection that was for a while my pride and joy. My own kid's collection, now scrimped from London parks and commons, continued to grow and I began to show it off to any passing uncle or grown-up in the hope of dunning a few bob off them. It was family holidays that kept the craze going and the willingness of Dad to indulge us. Indeed, though never managing to learn very much about butterflies and their habits, he could share our enthusiasm for the field. Like most fathers, he enjoyed being out and about with his boys, and, of course, he took the leadership role as his due. He fancied he had an eye for good ground, for the best spots. Chris and I would follow and, whenever we found butterflies, which was certainly not always, Dad would modestly claim that he'd known as much all along.

The climax of our outings with net and gas-bag came when the family moved from London to Germany and we took our first spring holiday in the south of France. I had looked forward to it as a lover desires his beloved, as a parched winter land longs for the spring. In my hard boarding-school bed I dreamt of myrtle-scented hillsides drenched in sunshine, of pursuing alpine butterflies in *The Sound of Music* settings while the doe-eyed cattle clonked their bells in the valley below. Each pinned and set butterfly would encapsulate moments from those precious days of freedom when nothing mattered but the allure of wonderful insects that the poets called sky-flakes or winged flowers.

That first spring day in the Alpes-Maritimes was magical. It had poured with rain all night, bouncing off the tin roof of the rented caravan like hail, but at length the cacophony slowed to a steady tap and then stopped. By morning the sun was shining through the blinds with what Nabokov called 'a long

glint of dewy brilliancy'. While the parents fussed over breakfast, Chris and I went out to see what we could find. Beyond
the caravans and chalets was a sunny wall with a border of
weeds and when we got there it was simply teeming with butterflies. Many of them were new to me. One looked like an
Orange-tip except that where ours was white this one was bright
yellow, like the Provençal sun rising through saffron clouds.
Another had black chequers over snow-white wings with a
contrasting dappled underside as fresh and green as the first
tender leaves. There was also a funny little skipper darting and
settling with marbled wings on the bare earth as though
hoovering up the warmth. And we barely had time to absorb
these novelties before there was a tigerish flash and a Swallowtail
alighted on an umbel of fennel right in front of us. The experience was like that moment in *The Wizard of Oz* when the whole
film bursts from black-and-white into Technicolor. Did we rush
back, grab the nets and lay into the butterflies? I hope not, for
this was stuff that dreams are made of. I could have watched
them for hours, or rather I could now; back then I might have
been in more of a hurry.

By the time we returned, breathless and bedazzled, our
mother had breakfast ready and both parents looked happy.
Later that morning we had our first proper butterfly hunt in
Europe. Dad liked the look of a particular valley just beyond
the terrace where a *paysan* was trimming his vines. *Definitely*
a good spot, he informed us.

'What's French for can-we-go-over-there?' asked Dad. 'Just
shout "*Là-bas?*"' I advised. '*Le Bas*,' shouted Dad, thereby
turning a phrase meaning 'over there' into one meaning 'you're
contemptible'. The man looked up in surprise. '*Tais-toi*,' he

yelled back. 'What does that mean?' asked my father. 'It means it's OK,' I told him hastily. 'Let's go.'

Dad was right. The wild valley seemed to have attracted butterflies from miles around. There were Camberwell Beauties that seemed to be almost the size of bats gliding overhead. The scented turf was brightened by darting sapphire points of Baton Blues and soon we were greeted by the sunrise-dazzle of a super-Brimstone known as the Cleopatra. It was like discovering buried treasure, and some of these butterflies were as bright as jewels. 'This is the area,' shouted my dad. '*This* is the area.' Gosh, we were a happy family that day.

Hunting butterflies in Europe is different from England. For a start there are more butterflies. On home turf you might consider yourself lucky to find a dozen species. In Europe, in peak season, you might expect to see thirty or more, although we struggled to name some of them because there were as yet no modern field guides. There was a greater sense of exploration, of not knowing quite what you might find but reckoning that it would probably be good. I also liked the fact that butterfly hunting in the Alps was not without its small dangers. Agility was useful when chasing mountain ringlets on loose rocks or bounding down a vertiginous slope after an Alpine Clouded Yellow. In rural France the farm dogs eyed butterfly collectors as they would a juicy steak. Local reactions to our activities varied. In France they shrugged their shoulders; it was just another thing *les foux anglais* did. In Germany hatchet-faced old men sometimes shouted '*Verboten*' as you dashed past but we didn't care. '*Verboten kaput*,' we would shout back over our shoulders, knowing we could run faster than them. Once, while collecting in what is now Croatia, we were mistaken for

Germans and memories are long in those parts. For a moment things looked very *verboten* indeed until they realised we were English. The English were their friends. The man could not have been nicer: 'Please, please, come to my house. Take all the butterflies in my garden. Look, there is a blue one. It is for you. *Dobro! Dober dan!* Come again.' Being English was a social advantage back then.

After the glories of the Dinaric Alps and the Mediterranean spring, collecting butterflies back at home was never going to be quite the same again. Even then I had the feeling that this game was nearly up. I had begun to feel self-conscious about carrying a net, and also, perhaps, increasingly guilty about catching and killing such lovely, carefree things. It was not conservation that was tugging my conscience. The butterflies we caught were all common and we never took more than a few. Nor was it waning interest, for, despite the onset of puberty and the usual teenage distractions, I was as keen as ever. But I remember a small French fritillary, freckled like a kestrel, with undersides marbled with purple and rufous-red and shot with piercing circles of silver. It was *Boloria dia*, the Weaver's Fritillary. When I came to set my 'series' of them I realised that something more than life had departed from those tiny corpses in their triangles of tracing paper. In death, the subtle interweaving of colour and texture had begun to unravel. Their purple iridescence had faded leaving behind a simple pattern of brown and black. 'Sad brown' was the phrase the early collectors had used for sombre colours, and it seemed particularly apt now. For the first time I felt the pity of what I was doing. Perhaps this delayed epiphany was just part of the business of growing up. I collected plenty more butterflies after that, but never with the same relish as those glory days in the sun with my brother and my father. Innocence, I felt, had left the chase. I carried on collecting, and it was still fun, but it no longer felt right. And so, gradually, I stopped chasing butterflies and started chasing girls instead, like any normal teenage boy. And with a great deal less success, I might add.

To a collector a specimen is much more than a dead insect stuck on a pin. It is, naturally enough, a desirable object, but also one with potential scientific value. In the past, many butterfly collectors specialised in unusual forms they called 'aberrations': a butterfly which had dashes instead of the usual dots, for example, or one where pale markings had been replaced with black. The challenge was to find these strange forms amid hundreds of ordinary butterflies and to locate places which contained more of them than others. Those places were often a closely kept secret, for mutual generosity among collectors only went so far. This was classic rarity hunting, the sole goal of many collectors, but such aberrations were also of great scientific interest. They offered clues to the way natural selection operates in the field, year on year, and from place to place. Collecting 'abs' led biologists such as E. B. Ford and Bernard Kettlewell to work out the implication of these alternative wing patterns and their relevance to the study of genetics – that is, the way different characteristics are passed on from one generation to the next: blue eyes or blood groups in humans, perhaps; extra spots or darker colours for butterflies. Hence, long after the death of the collector, old collections can still be used to produce new and original science. Today the national butterfly collection is being examined for the evidence it might provide on climate change, another area of study never envisaged when those butterflies were on the wing.

Some collectors bagged butterflies in much the same spirit as they shot selected game or angled for the biggest pike in the lake: as trophies of the field. Purple Emperors have a peculiar fascination for this kind of collector, exemplified by Ian Heslop (1904–70) who was as adept at catching high-flying Purple Emperors in English woods as he had been at shooting big game as a district commissioner in Nigeria. His collection, now housed in Bristol City Museum, contains no fewer than 185 Purple Emperors, many of them caught with his special 'high net' on the end of a twenty-foot pole. For him collecting had clearly gone beyond the usual bounds of the hobby and become an obsession. One of his diary entries reads: 'Holding the pole at the maximum extremity and with my hands at full stretch above my head, I edged the rim of the upper end of the net towards the Purple Emperor and then, with an upward jerk, struck . . . Time 4.14 p.m . . . probably the best shot of my "big-game" entomological career.' Whether shooting or netting, it was the same thing: a field sport. To his credit, Heslop watched living butterflies as well as amassing dead ones. His book, *Notes and Views of the Purple Emperor* (1964), has been called the first ecological study of any British butterfly.[2]

No doubt Heslop loved showing off his butterfly trophies. One of the pleasures of collecting of any sort is of sharing the passion with like-minded enthusiasts and also in competing with them. A fine collection offers the same pride and ego lift as a big car or membership of a smart club. But with butterfly collecting another, more generous spirit was at work: that of education. The eminent Victorian entomologist Henry Stainton held 'at homes' when visitors, especially young enthusiasts, were invited to browse through the collection while the great man hovered near, and also to bring with them insects, alive or dead, for identifying or exchange. A century earlier, the Duchess of Portland made her natural-history collections available on loan to artists and specialists. Through her amassment of valuable objects that others could study and publish she became, in effect, a patron of science. Whatever might have motivated such collections initially, the result was increased knowledge and greater understanding. And it must also have added its mite to human happiness.

In my own far more modest case, collecting benefited mainly a single person: me. A boyhood spent chasing butterflies opened my eyes to the natural world and eventually led me into nature conservation. Though I have long since lost interest in what is left of my collection, my boyhood obsession still reminds me of days spent among the butterflies and nights with the moths, and those I do not regret. Many of my butterflies were bred rather than caught and they remind me of different things: not of chases with a net but of searching for eggs or caterpillars; of gathering plants from the wild to grow on in pots or keep fresh in water for the ever-hungry caterpillars; of witnessing the transformation of the full-grown caterpillar to a seemingly inert chrysalis or pupa, followed by the always exciting moment when the adult finally emerges from the cracked chrysalis, dries its wings and reveals its full fresh beauty. Caterpillars grow fast, often changing into different forms as they do. And they are as different from one another as the adult butterflies, from the sleek, horned caterpillar of the Purple Emperor to the svelte grub of the Holly Blue, half-buried inside a berry. And every time you find the tiny egg of a butterfly or its well-hidden larva you get a small thrill of achievement.

My most valuable pinned butterfly is an English Large Copper. It was not caught by me; in fact I do not know who caught it or when, but it must have been a long time ago because the English Large Copper has been extinct for 150 years. It was given to me by a friend. My copper is a male, as copper-red and gleaming as the distant day when it flew over the still undrained fens or maybe was collected as a caterpillar and bred on (for there was a brisk trade in copper larvae). It is set with sloping wings in the fashion of the day on a contemporary black-enamelled pin with the top made of a ball of fine wire. Its posthumous history is outlined on three sets of data labels. It was already old when, the first label informs me, the object was sold at auction in 1902 to someone called P. Crowley. He might well have paid ten or more guineas for it, the going rate at that time when butterfly collecting was at its height. The second label tells me it was auctioned again in 1961 as Lot 303 by which time it sold for only two pounds ten shillings – a measure of how much less popular butterfly collecting had become by the mid-twentieth century. In due course it passed into the collection of my friend and then to me, perhaps its fifth or sixth owner. My contribution to its history was accidentally to snap the tip of one of its antennae. One day I will offer it to a museum. Precious English Large Coppers, of which there are only a couple of thousand surviving, should be in safer hands than mine.

What happens to butterfly specimens when their owners die or grow bored with them? In modern times the answer is often depressingly simple: they get thrown away – though the cabinet that contained them may be rescued. Probably only a minority of collections have long outlived the collector, most notably those important enough to be accepted by an institution such as a school or a museum. The financier Charles Rothschild spent much of his teens and early manhood acquiring a magnificent collection of all the swallowtail and birdwing butterflies then known. But on finding that completion can mean the end of one's interest, he gave it away to his old school, Harrow. Later on, combined with another large and valuable collection, it became the Rothschild-Peebles Collection, one of the finest in private hands. But most schools today find they no longer have much use for dead insects. In this case the cabinets were consigned to a cellar and, in 2009, the collection was broken up and sold. The way we study biology has moved on and what would once have been valued and curated properly has now become something of a millstone; with scientific potential still but deeply unfashionable and even faintly embarrassing.

So museums do not necessarily jump for joy when offered another consignment of pinned insects. For one thing, preservation is more difficult than before. Health and Safety legislation has banned the use of preservative chemicals such as naphthalene to keep the devouring mites at bay. The only remaining method is to lower the collection into a freezer, one drawer at a time. That increases the workload of already hard-pressed staff, and budget cuts are reducing the time available for curation. The result is neglect. The collections will suffer the fate of other exhibits whose time has been and gone, consigned to some storage space and slowly reduced to dust by museum beetles; in Professor Beth Tobin's phrase, 'suspended somewhere between memorabilia and rubbish'.[3]

In a sense that is understandable. Although a collection may have potential value to science, all of its personal load of memories and anecdotes will have been lost when the collector passed on. After the death of the Yorkshire naturalist and photographer George E. Hyde, in 1986, his widow donated his large and important collection of British butterflies and moths to Doncaster Museum. There, for a while, it was in safe hands, for the museum's keeper of natural sciences was the late David Skidmore, a leading entomologist who left his own collection of some 15,000 two-winged flies (Diptera) to the museum. But today the museum's staff are busy with other priorities. Several drawers of the collection are reportedly infested with beetles and at risk are Hyde's valuable collection of long-extinct English Large Coppers and Large Blues.[4] Doncaster's experience is probably typical of many local museums and institutions. At one field centre I visit regularly I have watched as the cabinet of butterflies and moths slowly crumble into dust leaving broken bits and labels stuck on bare pins.

The safest butterfly collection in the land is the national one, housed in the new Darwin Centre at the Natural History Museum. It is called the Rothschild-Cockayne-Kettlewell collection, or the RCK, after the names of the three famous collectors who left their specimens to the museum in the 1930s and '40s. Arranged in 5,000 large, glass-fronted drawers, it 'displays variation in all its forms', whether genetic or geographic.[5] For security reasons, viewing the collection is by application only, although some of it is now being made available online. I was once privileged to inspect the RCK, composed of a seemingly endless succession of polished mahogany cabinets known as Hill Units. It offers a universe of butterflies, as numberless as stars in the sky. I remember some of the highlights: extreme aberrations, all-black Purple Emperors; freak butterflies that are male on one side and female on the other; butterflies from the early days of collecting, bearing the names of famous entomologists; legendary butterflies one is slightly surprised to find actually exist. You view these marvels of nature from a standing position under inset ceiling lights in the eerie silence of the storeroom. The effect can be numbing. After a while there seemed to be pinned butterflies behind my eyes as well as in front of them. Quite soon I felt the need to retire for a cup of strong museum coffee, and it felt like coming up for air. I had the impression that visitors like me were quite rare; it may be different now that the collection has been rehoused in a new and customised building.

Lord Rothschild once claimed that among all his countless butterflies and moths, there were no 'duplicates'; not one. None was for swapping or exchange. To him the world's largest butterfly collection was the bare minimum necessary. Every specimen had significance. Rothschild was, of course, a very unusual man. He was an amasser of fine detail that was never-ending, and became an end in itself, circles within circles, ever more rarefied until he reached fine distinctions that only he and his curators could appreciate. This, you sense, is the ultimate frontier of collecting, the polar opposite of a boy's day out with a net and a jam jar. It is magnificent but it can also seem like an enormous but empty monument, a vast mausoleum of former life. In a changed era, Rothschild's life's work slumbers under the lights, a million dead butterflies in their scented drawers. The rainbow essence of a hundred Victorian summers.

3.

Graylings

The Birth of a Passion

They say humanity divides into two kinds of people: those who want to know where something has come from and those who are more interested in how it works. In the last chapter I described how butterfly collecting 'worked'. In this one I will try to pinpoint where this passion for butterflies came from and what sustained it for so long.

We could find a beginning in our oldest butterfly. Where is it? Dead butterflies are fragile things and yet, properly curated, they can last a surprisingly long time. Professor E. B. Ford of Oxford thought the oldest butterfly was probably a tattered Bath White, in Oxford, naturally enough. According to its label it was captured in May 1702 and it was passed, miraculously intact, from one collection to another until it ended up in Oxford's University Museum. Back in 1702 the butterfly was known not as the Bath White but Vernon's Half-Mourner, and it had no Latin name because they had not yet been invented. But Ford rarely paid much attention to the world outside Oxford and he was wrong about this being the oldest butterfly. He was out by more than a hundred years.

My own candidate is a flattened, faded wisp, almost the ghost of a butterfly, found preserved in the leaves of a manuscript dating back to 1589. With touching appropriateness, the manuscript was that of the famous *Theatre of Insects* of Thomas Moffet, the first book about insects, at least from Britain. Moffet was a Londoner, a Fellow of the College of Physicians and, at least in a Tudor sense, a naturalist. As a medical man he was interested in the using of ground-up bugs in various potions though he made an exception of 'Day Butterflies' (as opposed to 'Night Butterflies' or moths). These he admired for their own sake, for their beautiful 'colours, attire, rich apparel, roundles, knots, studs, borders, squares, fringes, decking [and] painting'.

Moffet's butterfly was discovered during a restoration of the original manuscript of the *Theatre* by the British Library. Though much faded, and minus its head and body, it is still easily recognisable as a Small Tortoiseshell. Evidently Moffet had taken some care over it for the wings are symmetrical and spread as in life. It is, in fact, a very good match for Moffet's illustration of the butterfly, a woodcut based on an original watercolour. Could it be the very specimen that prompted Moffet's rhapsodic description of the Small Tortoiseshell? One that celebrated its 'light blood-color, dipt with black spots [with] golden crooked lines like the Moon, being itself a murry, nicked on the sides like a saw'.[1] Moffet's house, as we know from his own account, was full of cobwebs. Perhaps the butterfly had slipped through the window casement, as tortoiseshell butterflies do in late summer, seeking some cool, dark corner to hibernate, only to end its short life in a spider's web.

For the oldest known butterfly *collection* we need to fast-forward nearly a century to the England of Leonard Plukenet (1642–1706). Professor Plukenet was a distinguished botanist, the Queen's gardener and one of the first Fellows of the Royal Society alongside Isaac Newton and Robert Hooke (he is remembered in the genus *Plukenetia*, a low bush from South America with star-shaped fruits known as Inchi or the Inca-peanut). Plukenet's butterflies are among a collection of insects he made by simply glueing their bodies on to the pages of a blank-leaved book. Among those that still cling to the page more than 300 years later are various squashed tortoiseshells, Peacocks, Meadow Browns and Brimstones. Plukenet's 'insect book' was overlooked until recently because it was accidentally included in the botanical collections inherited by the Natural History Museum and not among the insects. The disadvantage of pressing and glueing insects into a book, of course, is that every time the volume is opened, bits of brittle, dried insect dislodge and fall out. Patches of sepia-toned paper now mark many of the places where Plukenet's butterflies have dropped out over the years. His book is now kept in an airtight cabinet and only opened on special occasions.

Almost as old is another collection in the museum that not only pressed butterflies like flowers but pasted them in *with* pressed flowers. It once belonged to Adam Buddle, a clergyman and botanist who lent his name, with unconscious appropriateness, to the Butterfly Bush or *Buddleia*. His butterflies, frozen in mid-flutter amid sepia-toned rushes and other plants, prefigure the paintings of later artists who liked to show butterflies and flowers together. Buddle was careful to pen a short description of each one in Latin so that, even when the butterfly is missing, or represented now by only a sliver of leg or a patch of glue, we know what used to be there.

The oldest properly mounted butterflies once belonged to a London apothecary and Fellow of the Royal Society called James Petiver (1663–1718). Perhaps copying the insect collections he had seen in Amsterdam, he mounted his dead butterflies on a pin and spread their wings into a natural posture on a board much as collectors did two centuries later. But contemporary pins were thick with big round heads and, being made of soft iron or tin, were apt to bend or corrode. They were suitable for a butterfly the size of a Red Admiral but not for smaller skippers and blues, and still less for tiny moths. Petiver found an alternative method by carefully spreading the wings and holding them in place between two slivers of mica, a transparent natural material that resembles clear plastic. Once labelled and sealed with strips of gummed paper the result looks remarkably like a slide mount with a real butterfly in place of the developed film. One advantage of this method was that the butterfly 'slides' could be stacked together neatly in a cheap box. It also enabled you to handle the specimen safely and examine both sides of the wings, an advantage denied to glued butterflies. A few of Petiver's mica-mounted butterflies still survive in the Natural History Museum. One that I once held in my hand is the first-known specimen of a Brown Hairstreak, caught, so the label tells us, by Petiver himself at Croydon in August 1702.

An awkward problem for these pioneers was how to kill the butterfly without harming its delicate wings. Beetles and spiders could be drowned in spirit but a butterfly or a moth needed more careful handling. Petiver's first solution was crude and, one would have thought, impractical. He put his butterflies to death 'by thrusting a pin [in] their body and sticking them in your hat'. He also advocated 'gently crushing their head and body between yr fingers which will prevent their fluttering'.[2] Pinching the thorax of a butterfly and bursting its heart results in almost instant death. The novelist and lifelong butterfly collector Vladimir Nabokov used this method all the time. But success requires nimble fingers and long nails.

Eighteenth-century butterflies died a variety of ugly deaths, some by a prick of a fine pen nib dipped in *aqua fortis* (nitric acid), others suffocated in the fumes of burning sulphur or by steam from a boiling kettle. One collector despatched his insects by holding them with forceps over a burning candle, at some risk, one would have thought, of the bug catching fire. Ether was apparently used to kill butterflies long before it became a widely used anaesthetic. By the nineteenth century many collectors had turned to cyanide, either produced naturally from crushed laurel leaves or, far more dangerously, in lump form, bought from a friendly chemist. Noxious fumes from these various chemical substances eventually replaced more home-made methods of killing. The butterfly succumbs to what looks like paralysis in a few seconds with unblemished wings.

No one seems to have speculated about whether butterflies suffer pain. Many come to atrocious ends in the wild, stuck in a spider's web, nibbled alive by mites, or suffering a lingering death from a parasite. To feel pain, philosophers assure us, you require a capacity for emotion. Is the insect world emotion-free? One possible answer was suggested by George Orwell who, in a cruel moment, cut in a wasp in half with a knife as it feasted on a puddle of jam. The wasp just carried on eating the jam. It wasn't going to let a little thing like the severance of its lower body interfere with a necessary metabolic process.

In due course Petiver's butterflies passed into the greatest collection of natural objects of the day – the one that kick-started the British Museum. This was made by the wealthy philanthropist and antiquarian Sir Hans Sloane, who displayed his trophies in a magnificent gallery 110 feet long, crammed with cabinets of shells, corals, crystals, bird's eggs, skeletons and skins. Among them were thousands of 'brilliant butterflies' gathered from all over the world by the crews of merchant ships keen to augment their wages by procuring specimens for collectors.[3] Sloane was not the only one. Dru Drury, a London goldsmith and a connoisseur of insects, spent thousands of pounds on his worldwide collection of butterflies and engaged artists to paint likenesses of the best ones (the collection has long gone but the paintings remain). The Danish scholar Fabricius viewed Drury's butterflies 'with as much glee as a lover of wine does the sight of his wine cellar when well-stocked with full casks and bottles'.[4]

What happened to all these butterflies from the Age of Enlightenment? The national collection is full of Victorian and Edwardian butterflies but it has relatively few from before the turn of the nineteenth century. Perhaps that will seem less surprising when we recall what happened to James Petiver's butterflies. Petiver, it seems, was not a tidy man. According to one visitor, his celebrated museum, the *Musei Petiveri*, looked like a junk shop.[5] When his butterfly collections passed in due course into more careful hands they were already in a sorry state. Sloane, who paid £4,000 for Petiver's hoard after the latter's death in 1718, confirms the Hogarthian conditions in which the objects were kept:

> He had taken great pains to gather together the Productions of Nature in England; but he did not take equal Care to keep them, but put them into heaps, with sometimes small labels of Paper, where there were many of them injured by Dust, Insects, Rain etc.

Sloane paid tribute to Petiver's 'industry, though not to his tidiness'.[6] Yet he held his eccentric friend in high regard and was one of the pallbearers at his funeral.

In Sloane's great gallery the collection, or what was left of it, was well looked after. To preserve butterflies from the ravages of mites, collectors of his day rubbed each box or drawer with 'Oil of Spike' or lavender oil – which must have lent them a pleasant scent, unlike the harsher smells of camphor and naphthalene in later times. But when Sloane himself died in 1753 leaving his collections and objects to the new British Museum, much weeding-out took place. Museum curators, now as then, do not automatically welcome old collections and, indeed, may see them as more of a burden than an asset. The man in charge of the British Museum wanted new, topical and exciting objects to display and enhance the institution's prestige, not a pile of old junk from the past.

And so, when Parliament decided to investigate the state of our first national museum in 1835, it seems that there was very little left of the first great international collection of butterflies. This is what Charles Koenig, the museum's then 'Keeper of Natural History and Modern Curiosities', had to say about it:

Questioner: 'Is the entomological collection which was left by Sir Hans Sloane in a perfect state at present?'
Koenig: 'There is hardly any of it remaining.'

'How does it happen that the collection has been lost?'

'When I came to the Museum most of these objects were in an advanced state of decomposition, and they were buried or committed to the flames one after another. Dr Shaw [the previous curator] had a burning every year; and he called them his cremations.'

'Is there any single insect remaining of the 5,439 which were presented by Sir Hans Sloane?'

'I should think not.'

'Do you think that so great a destruction of specimens can solely arise from Natural Causes?'

'We considered them as rubbish, and [as] such were destroyed with other rubbish.'[7]

Hence what survives of our oldest butterflies did so by accident: Adam Buddle's pressed butterflies, for example, escaped because they were accidentally stored among botanical specimens in the herbarium instead of in the insect gallery. It is thanks to such oversights that we still have a few faded butterflies to bring us face to face with the origins of entomology: heirlooms from the countryside of Queen Anne.

What created this apparently sudden interest in butterflies around 1690? Until James Petiver began to collect and draw them, it seems that butterflies did not even have names. At any rate, none of the first books to illustrate butterflies, by Thomas

Moffet, or Christopher Merret or Martin Lister, had any names for them. Yet by Petiver's death in 1718 a fashionable interest in butterflies and moths had grown up – sufficient at any rate for the artist Eleazar Albin to find a long list of aristocratic subscribers for his expensive art book on the subject. And this time they did have names; not necessarily those we call them by today but good English names that everyone could use. People, evidently, had started to talk about butterflies.

Why was that? Was it increased leisure? Or the availability of coffee houses where people could meet and, stimulated by caffeine and tobacco, talk their heads off? Perhaps one answer is that more men and women than before were reading and thinking for themselves. They had broken through the 'received wisdom' of the Church. The works of Francis Bacon (1561–1626) were popular and people were impressed by his portrayal of an ideal society dedicated to discovery and the pursuit of knowledge. When the Royal Society was formed by Royal Charter in 1663, its declared purpose was 'to improve natural knowledge' by discussion and experiment. So one of the great discoveries of the Enlightenment might have been intelligent conversation. Quite suddenly, it seems, the natural world was found to be interesting. Perhaps the best word for it is 'biophilia', Edward Wilson's new-minted word meaning a love of nature. By taking an interest in the natural world not simply as it was presented to them but through observation and method, people found that nature engaged their feelings as well as their minds. With that came a desire to incorporate the 'productions of nature' into their homes in the form of gardens, menageries and collections. For the great English naturalist John Ray (1627–1705), 'biophilia' took a particularly exalted form in a wish, far ahead of his time, to look for order amid the apparent chaos of nature. For lesser mortals it took the form of the curio cabinet. While few could aspire to the wisdom of Bacon or the curiosity of Ray, anybody with a minimum of resources could collect shells or fossils – or even butterflies.

Butterflies were pretty, they could be captured easily enough and, unlike some insects, their colours did not fade after death. People were also intrigued and charmed by their complicated life stages, and especially by the seemingly miraculous 'transformation' from caterpillar to winged butterfly. But at the beginning there was nothing to help them. There were no clubs or societies where experience and knowledge could be shared. There were no identification guides, indeed no worthwhile books at all unless your pocket could extend to leather-bound volumes in French or Dutch. The would-be English naturalist was hardly better off than a castaway on an unknown island.

What seems to have got things going was John Ray's plan, in the 1690s, to write a book about the natural history of insects to match his published volumes on animals and plants. This was necessarily a long-term project and by then Ray was past his prime, getting on in years and in poor health. Fortunately he could count on the help of his family and neighbours, and on a wider circle of educated acquaintances in Oxford and London, as well as certain rural schoolmasters and fellow members of the cloth. Such, for example, was John Morton who caught him a White Admiral 'in Essex not far from Tolbury' in 1695. Another was the Reverend Mansell Courtman who hit the jackpot that same year by capturing a 'large black butterfly with wings marked with white', probably a female Purple Emperor.[8]

Some of Ray's helpers were men of standing. Among them are a famous botanist (Plukenet), two physicians (Samuel Dale and David Kreig), a distinguished garden designer (Tilleman Bobart), a silk-pattern designer and artist (Joseph Dandridge), a silk trader (Charles DuBois), an apothecary (James Petiver), a clergyman (Adam Buddle) and an Eton schoolmaster (Robert Antrobus). Apart from Ray himself, at least three – Plukenet, Petiver and Kreig – were Fellows of the Royal Society. Many of them were botanists for whom butterflies had an understandable fascination. Ray himself was primarily a botanist. So was Petiver, whose neglected day job was to manage the physic garden of the Society of Apothecaries. A knowledge of plants and gardening was useful when it came to searching for the eggs and caterpillars of butterflies and rearing them on plants in pots or in water. Work like this turned them into observers while their heightened sense of curiosity made them naturalists, even before that word had been invented.

With so many like-minded enthusiasts about, it was only a matter of time before some of them formed their irregular meetings into something more substantial. Being Englishmen they did what came naturally and formed a club – the first entomological society in history. But because entomology then barely existed as a science, they called themselves something else. They became the Society of Aurelians.

During his years of exile in Berlin in the 1930s, Vladimir Nabokov wrote a short story about a dealer in butterflies. This dealer was down on his luck. His life felt empty and boring; his work didn't interest him and his marriage was on the rocks. His one dream of escape was to go abroad on a collecting trip, but he couldn't afford it. In the end he managed to finance an expedition by swindling a customer but on the day of departure he suffered a fatal stroke. Even so, the all-knowing narrator informs us, his life was not a complete failure. The man had reached a state of happiness where his inner world could compensate for the wretchedness of his earthly existence. In his mind he was already in those far-off lands finding 'all the glorious bugs he had longed to see'.[9] The man's inner journey echoed the metamorphosis of insects, taking him from an inert, pupa-like condition to the transcendent state symbolised by the butterfly.

In typically cryptic fashion, Nabokov chose to call his tale 'The Aurelian'. Aurelian is not a familiar word today, even among entomologists. Its literal meaning is 'the golden one'. The analogy is with the chrysalis – another word meaning gold – of Red Admirals and certain other butterflies whose pupal cases are dappled with gold or silver spots. By calling themselves Aurelians these Georgian naturalists were making a form of self-identification with butterflies. We do not know exactly when the Society of Aurelians was formed, for all the records are lost, but it was flourishing in London in the 1730s and '40s. It met at the Swan tavern at Cornhill in the heart of London on a narrow passageway thronged with booksellers and coffee houses. Many of its members also met at the nearby Temple House where they drank their coffee and talked about the natural world.

The Aurelians were all men; in that age women were not permitted to be members of gentlemen's clubs although there were certainly women who were interested in butterflies. The names of some of the members were listed in a later book about butterflies and, interestingly, the best-known of them were not 'scientists' but artists, designers and poets. The president and probable founder was Joseph Dandridge (1664–1746), a leading silk pattern designer and amateur artist. He was among the first to paint realistic images of birds, insects and spiders and was much admired by his fellow Aurelians for his knowledge and expertise. Another leading member was Henry Baker (1698–1774), a self-made man and amateur poet who was also a pioneer in therapy for the deaf. His other passion was for the microscope. His book *The Microscope Made Easy*, first published in 1743, brought the pleasures of miniature worlds to prosperous London drawing rooms. He was a true man of the Enlightenment, a founder of the Society of Arts, a gardener credited with introducing rhubarb to Britain, and the sometime editor of a weekly, *The Universal Spectator*.[10]

Another prominent member was James Leman (1688–1745), also a London silk manufacturer and pattern designer, and a liveryman of the Company of Weavers. His textile designs borrowed from natural patterns in nature, particularly wild flowers, in a way that prefigures William Morris. Silk, of course, comes from a silkworm, the caterpillar of a moth. Leman's colleague and fellow Aurelian, Peter Collinson (*c*.1693–1768), was also in the textile business, a trader in flax, hemp, silk and wine. It was on one such trade-promoting journey to the New World that Collinson made a pioneering collection of American butterflies. Later on, this wealthy and long-lived Quaker corresponded with Linnaeus over the ordering of insects.[11] His collection was one of those purchased by Sir Hans Sloane and hence left to the British Museum – and to the awaiting bonfires of Dr Shaw.

Thomas Knowlton (1691–1781) represented another

prominent Aurelian vocation: he was a garden designer. Over his long career he created important gardens in Oxford, at Eltham in Kent and at Canon's Park at Harrow, now buried under suburban streets, as well as the still-extant garden of the Earl of Burlington at Londesborough in the East Riding. Knowlton worked for William Sherard, Oxford's Professor of Botany and very likely his interest in wild plants led to a similar interest in butterflies and other insects.[12] He was probably an associate of Tilleman Bobart, another garden designer who preceded him at Canon's Park and who supplied butterflies and moths to both Petiver and Ray.

It seems to have been the usual practice for all these early collectors to donate specimens to the common collection, held, along with the society's books and documents, at the Swan. It deeply impressed the young artist Benjamin Wilkes with its 'Specimens of Nature's admirable Skill in the Disposition, Arrangement and contrasting of Colours'. Wilkes himself was admitted to the society and it is mainly through his eyes that we see the Aurelians at work and at leisure in their London setting. He praised the society and its works in the introduction to his master work, *The English Moths and Butterflies*, published in 1749 – shortly after the society had ceased to exist.

It came to a premature end in dramatic fashion. On the night of 25 March 1748, a fire broke out at a periwig maker's in Exchange Alley and quickly swept through the Cornhill district of the City. A hundred houses went up in flames, catching people asleep in their beds, as well as revellers in Cornhill's many public houses. The Aurelians were enjoying a late-night sitting at the time, perhaps all slightly drunk after dining at the tavern. It was New Year's Eve by 'Old Style' reckoning – a night for sitting up late. Years later, Moses Harris recalled what happened next:

> The Society was then sitting, yet so sudden and rapid was the impetuous course of the Fire, that the Flames beat against the Windows, before they could get well out of the Room, many of them leaving their Hats and Canes; their Loss so disheartened them that altho' they several Times met for that Purpose, they never could collect so many together as would be sufficient to form a Society.[13]

And that was the end of the world's first entomological society. There was no time to save the society's books, records or collections, nor what was intriguingly referred to as its 'Regalia'. The entire archive and collection was consumed at a stroke. At length a second and then a third Society of Aurelians sprang from the ashes of the first, but neither lasted very long. Internal dissension and apathy – perhaps the novelty was wearing off – caused both of the new societies to fold inside a few years. Their direct successor in the nineteenth century was the more modern-sounding Entomological Society of London. The change in name symbolised a deeper cultural change. In the eighteenth century butterflies had belonged to the artists: engravers, watercolourists, pattern designers, poets. In the nineteenth and henceforth, the study of butterflies was taken over by amateur scientists. They had become part of the new science of entomology.

Meanwhile butterfly collections were getting bigger and bigger. In the early days most collectors had been content with a short series of butterflies of each species: a couple of males and a female, and perhaps one mounted upside down to show the undersides. But by now the fashion was for a long series, perhaps even an entire drawerful of the same species. Collectors rarely explained why they wanted so many examples. One obvious reason, common to all kinds of collecting, was peer pressure and competition: the biggest collections were seen as the best. It was comparable to a game bag – and some collectors of the Victorian era were as fond of shooting grouse and pheasant as they were of pursuing butterflies and moths. Another reason was that Victorian collectors were obsessed with 'varieties'; that is, forms of a butterfly whose wing patterns differed from the norm. Naming these became a kind of pseudo-science, and one collector, Henry Leeds, invented a complicated method of names which amounted to a new language, including such tongue-ticklers as 'ab. *Infra-semi-syngrapha-grisea-lutescens*' – which must have needed a label bigger than the butterfly![14]

Variety collecting was sometimes derided as the entomological equivalent of stamp collecting: simply amassing different kinds for the sake of it. It was only in the twentieth century, with the discovery of genetics, that the significance of all these 'aberrant' forms began to be understood. Professor E. B. 'Henry' Ford (1901–88) had been a keen butterfly collector from boyhood. With his father, he kept a colony of Marsh Fritillaries, a butterfly which displays exceptionally variable colours and patterns, under close observation for nearly twenty years. He discovered that the butterfly's numbers fluctuated from year to year and that the varieties were most frequent when it was enjoying a good year. This provided Ford with a moment of supreme insight. Variation in wing pattern allows a species 'to adjust itself to its environment more rapidly than it would otherwise do, and enables us to examine actual instances in which an evolutionary change takes place in nature'.[15] In other words, butterfly varieties showed how evolution works, not by slow, imperceptible degrees as Darwin had thought, but right in front of our eyes. Ford's work on butterflies revealed that evolution is happening all the time, out there in the woods and fields. The genes that govern the patterns on a butterfly's wing are passed down from one individual to the next, generation after generation, through interactions that we call genetics. A species is constantly readjusting itself to its environment: to changing weather, changes in predation or changes to its habitat. Without this year-by-year capacity to adapt, no species could survive for long.

Of course the butterfly does not 'know' this, any more than we knew it until Henry Ford came along with his butterfly net. But the butterfly's genes 'know' it. The transference of genes by inheritance, from one generation to the next, is the function of all life on earth. Every species, from a Marsh Fritillary to a blue whale, is the custodian of its genes, the immortal part of every species including our own. It is fitting that butterflies were one of the conveyers of this fundamental scientific truth. For butterflies have been seen by some, from the ancient Greeks onwards, as the poetic equivalent of genes: the embodiment of the soul. I will return to this topic in Chapter 9.

Amassing a full range of variation was therefore another reason why butterflies were collected in large numbers – though doubtless most collectors were more interested in the rarity of 'varieties' than in their contribution to science. Fortunately collectors could and did share their observations and exploits in one of a thriving number of journals: the *Entomologist's Annual*, the *Entomologist's Monthly Magazine* or the *Entomologist's Weekly Intelligencer*. A twentieth-century collector, Baron Charles de Worms, routinely described his entomological travels which, in 1937, took him on an 'on the whole successful tour of the British Isles'. His field attire was typical of many insect collectors: an old tweed jacket, a couple of tattered old pullovers and a satchel bulging with nets and boxes. When chasing the Purple Emperor he carried a pocketful of ripe blue cheese which he would spread over gateposts in the hope of luring the butterfly, notorious for its love of smelly products, down from the treetops. An earlier generation sometimes wore cork-lined top hats in the field. They were useful receptacles for pinned insects. One collector even absent-mindedly kept his bottles and boxes in his hat, only to have them spill out whenever he raised it to a lady.

Victorian collectors were obsessed with the appearance of their butterflies. They must be set with mathematical precision and without the smallest blemish, and, in the case of moths, with two little forelegs poking out as though the moth was about to crawl away. They were displayed in rows like soldiers on parade with a thread of black cloth dividing one line from the next. Sometimes, to save on space, the butterflies overlapped one another so that the effect was not so much a row of individuals as a coloured river of wings and scales running from the camphor cell at the top of the drawer to the brass handle at the bottom.

Only fresh, perfect specimens were admitted to the cabinet. That meant, of course, that most butterflies were perfectly safe from the collector's attentions since their wings become faded and tattered very quickly. Gone was the Georgian idea of arranging butterflies in geometric or kaleidoscopic patterns (which strangely prefigured the butterfly work of Damien Hurst). That kind of frivolity would have appalled a Victorian collector who was, above all, deeply serious.

Many nineteenth-century collectors regarded themselves as scientists and without a reference collection there could be no science worthy of the name. Conservation was not then a concern even though the morality of collecting large numbers of rare butterflies was occasionally raised in entomological journals. That issue grew and became polarised in more recent times in an argument epitomised by two novelists with sharply differing opinions on the subject: John Fowles and Vladimir Nabokov. But before rehearsing what their argument was about, let us go hunting butterflies in a much earlier era: out into the fields of medieval Brabant in the company of the biographer of the Black Prince, Jean Froissart.

4.

Gatekeepers

Collecting with Jean Froissart,
John Fowles and Vladimir Nabokov

Among the treasures of the British Library is a 700-year-old
illuminated manuscript called *The Romance of Alexander*. It
dates from the 1330s, at the start of the Hundred Years War,
and, appropriately enough, it is about the archetypal war leader,
Alexander the Great, one of the 'Nine Worthies' in the medieval canon of heroes. But what makes this particular document
so remarkable is not the story it tells so much as the little pictures
that the illustrator has inserted into every available space. Few
of them seem to have much to do with the exploits of Alexander.
On one page there might be a trio of dancing monkeys – a
fourth plays a lute – while on another some wild, hairy men
from the far side of the world are grimacing through faces
carved into their chests. Some of the pictures may have a satir-
ical purpose, such as the monk and nun playing leapfrog or the
giant hare which hunts men with his crossbow.

Among the dancing monkeys and leaping nuns are identifiable animals and birds, among them a goldfinch, a cuckoo, a great tit and a hoopoe. And there are butterflies, too, fluttering around the margins of the manuscript. Some are nondescript or imaginary, as big as fruitbats with big knobbly horns on their bug-eyed heads. But others are the butterflies that we know and love, among them Peacocks, Clouded Yellows and Red Admirals.

These 700-year-old butterflies are not shown merely for decorative purposes. People are trying to catch them. In one scene a fantastic butterfly with a long, worm-like body hovers over what looks like its future prison: an elegant pot. A youth seems about to sweep it up in his badly drawn net but has been distracted by a woman holding another jar and covering the opening with her hand. Nearby, six children, all huddled together, are watching intently and making gestures of desire. A different, equally lively scene shows a group of well-dressed women catching butterflies by seizing hold of their wings. What, exactly, is going on and why are all these men, women and children so eager to catch the butterflies?

One thing seems certain. They are having a good time. We can see that from the expressions on their faces. Even the butterflies seem happy for the illustrator has sometimes drawn little smiles on their horny faces. And those badly drawn children all huddled together in a heap are evidently anticipating some game or simple pleasure.[1] We know from Shakespeare that children enjoyed chasing after butterflies. Little Coriolanus tore them to shreds to show what a tough boy he was. But King Lear and his daughter Cordelia looked forward to simply laughing at 'gilded butterflies' as a simple pleasure in store for them as they whiled away their time together in captivity.

Perhaps the butterfly scenes on the Alexander manuscript are another satire, holding up foolish behaviour for derision. All the same these village folk from the England of the 1330s seem to have plans for their capture. I had no idea what they might be until I read Christina Hardiment's biography of Thomas Malory, the author of the fifteenth-century *Morte d'Arthur*. Reduced to guessing what Malory's childhood might have been like, Hardiment quotes from the *Chronicles* of Jean Froissart who was a young man at the time when unknown clerks were busy at work on *The Romance of Alexander*. Froissart provides us with a rare glimpse of a medieval childhood. He lived in Hainault, in what is now Belgium, but we can assume that his experience was not wildly different from that of children on the opposite side of the Channel. He enjoyed doing what every boy has done down the ages. He made mud pies and built dams across streams. He blew soap bubbles from a pipe. He played follow-my-leader and hide-and-seek and even a medieval form of charades.

But, more surprisingly, he enjoyed playing with butterflies. Young Jean liked to fasten a fine flaxen thread to their tiny living bodies, tie the other end to his hat and then let them go. The butterflies would thereupon flutter around his smiling face like tiny living kites or tethered elves. 'I could make them fly where I pleased,' recalled Froissart, and you are left to imagine the butterflies flitting gaily about him as he trod the byways of old Hainault.[2]

By way of confirmation, a second manuscript from the British Library, made at about the same time as *The Romance of Alexander*, actually shows young men flying tethered butterflies in this way. Perhaps they were bought from a character like the unforgettable 'Rose-beetle Man' in Gerald Durrell's *My Family and Other Animals* with his 'lengths of cotton to each of which was tied an almond-sized rose-beetle, glittering golden green in the sun'. If *The Romance of Alexander* had been a comic, the next picture might have been six happy children holding butterflies on threads, fluttering and settling in an endless merry-go-round of brilliant red, yellow and purple wings.

COLLECTING WITH JOHN FOWLES AND VLADIMIR NABOKOV

Spring 1875 was cold and damp, like a soggy flannel. It had been a rotten winter too. December was freezing, dropping to minus 14.5 degrees C. on the Suffolk coast on New Year's Eve. January was milder but sopping wet and February full of deep, stormy snowfalls. And then it got worse. Just as the buds were breaking and hibernated butterflies were emerging from their hiding places, a volcano in Iceland erupted and filled the sky with dust. May was wet, and so was June, and July the wettest month of all. But then, suddenly and unexpectedly, the clouds cleared and England enjoyed a belated summer. The sun shone for weeks, from the second week in August to the end of September. Those who could took a holiday. A certain Mr H. Ramsey Cox, for example, went to the Isle of Wight with a couple of friends to collect butterflies. He was in luck – stupendous luck. His outing coincided with an almost unheard-of mass migration of the very rare Pale Clouded Yellow. It was what collectors called 'a *hyale* year', after the butterfly's Latin species name, *Colias hyale*. Hundreds of pale yellow migrant butterflies struck land under the hot August sun, settling on thistles and laying on the blue pagodas of lucerne then sown in fields to grow fodder for horses. Ramsey Cox must have reasoned that luck like this was unlikely to strike twice. So he ripped into the pale yellow swarms, net in hand. Working methodically at what he referred to afterwards as his 'innocent pleasure', Cox and his friends managed to catch 800 Pale Clouded Yellows. It was almost certainly a record, though perhaps not one any normal person would be proud of.[3]

Even fellow collectors questioned the morality, even the sense, of slaughtering hundreds of a single rare species. 'What pleasure could possibly come from taking the lives of eight hundred defenceless *hyale*?' asked a contributor to the letters page of the *Entomologist*. What could possibly be his motive? Ramsey Cox replied unapologetically in the next issue. He admitted his 'success' and, yes, it was all quite true. It *was* 800 butterflies, more or less. But what was wrong with that? Some he intended to give away to friends, some he might swap and the rest looked very nice in his collection. No one was harmed by it, and there would be plenty more butterflies where they came from. At this point the editor stopped what might have been an interesting response by asserting that 'this little passage-of-arms must end here'.

Probably it was coincidence that the Pale Clouded Yellow also graced the dust-jacket of that classic novel of the 1960s, *The Collector*. Tom Adams's jacket design for John Fowles's novel shows a female Pale Clouded Yellow pinned on cork along with a tress of blonde hair and the sort of big key that locks the door of a cellar. The lead character in the novel, Frederick Clegg, is a far nastier piece of work than H. Ramsey Cox. Back in 1875 it was not collecting per se that had been stigmatised but *excessive* collecting, and even then not for reasons of conservation so much as fair play and sportsmanship. The word used both by Ramsey Cox and his detractors had been 'innocence'. But although Fred Clegg is evidently a virgin, he is no innocent. His obsession with collecting butterflies leads directly to a different kind of obsession with fatal consequences. In essence the novel represents a full-on attack on the mentality of collecting.

A robin redbreast in a cage, wrote William Blake, puts all heaven in a rage. Butterfly collecting seems have done that to John Fowles and he is not alone. One butterfly website not untypical of many others asks us: 'Why would you need to have a cabinet full of dead butterfly specimens?' before delivering the answer, and by implication, the *only* answer. It is because you, the collector, are 'selfish, greedy and obsessive'. If you spot anyone with a net, the narrator continues, ask them if they have permission. If you suspect they are collecting illegally, phone the police. Make their lives hell.[4]

This rage has grown up during my lifetime. The publication of *The Collector* in 1963, and the film based on it two years later, marked the approximate point at which public attitudes to specimen collecting began to change. Yet there are only passing references to collecting in *The Collector* which is not really about that hobby so much as the mentality that Fowles believed lay behind it. Clegg kidnaps an art student, Miranda Grey. The ensuing story is told from the perspective of both Clegg and the cultivated, middle-class girl he has locked up. Clegg wants her to fall in love with him, and relations with her loathsome captor do eventually improve at least to the point of pity. Miranda is of course the metaphorical pinned butterfly – Clegg even uses chloroform to catch her. 'I am one in a row of specimens,' she writes in her diary. 'It's when I try to flutter out of line that he hates me. I'm meant to be dead, pinned, always the same, always beautiful . . . he wants me living-but-dead.' For Fowles the butterfly is a metaphor for beauty and freedom, and once collected and pinned to a board it becomes beauty without freedom.

By collecting obsessively, like everything else he does, Clegg, in Miranda's words, 'hoards up all the beauty in these drawers' (i.e. the drawers of his cabinets). Once he has added Miranda to his collection, he gives up his raids into the Sussex countryside. But what little we see of his hobby is made as horrible as possible. Sometimes Clegg rips the wings off the butterfly or squashes it between his hands and wipes off the mess with grass. Collecting is the key to Clegg's sadistic nature. He is eaten up by an awareness of his own vileness. Even before he gets his hands on the girl, it is clear he is a potential psychopath. *The Collector* is a fairly nasty story, even nastier to read today, perhaps, than in 1963. Novelists should not be confused with their novel, but it is hard to read *The Collector* and not deduce that John Fowles hated butterfly collecting with a passion.

And if you did assume that, you would be right. One of the reasons Fowles gave for writing the novel, his first, and calling it *The Collector*, 'was to express my hatred of this lethal perversion'.[5] Fowles returned to the subject in a foreword for a booklet published in 1996 called *Collector's Items* written by a feminist writer, Kate Salway. *She* saw naturalist-collectors as retarded, cold-blooded and above all *male* predators of everything that is alive and beautiful. Her pages are flooded with photographic images that are framed deliberately to make the naturalist's tabletop look like a torture chamber. Pinned butterflies are the only bright colours among the cold steel of forceps and pins and dead grey of sheets of muslin. 'All natural history collectors,' noted Fowles, 'in the end collect the same thing: the death of the living.' We can never begin to understand nature, and certainly will never respect it until 'we disassociate the wild from the notion of usability'. To 'use' nature by collecting it, was, in his opinion, to defile it, and along with that, ourselves.

On the way to this bleak, schismatic conclusion Fowles admits that he had once been a keen collector himself. 'I began very young as a butterfly collector, surrounded by setting boards, killing bottles, caterpillar cages.' He also shot game birds in what he recalled later as 'a black period in my relations with nature'. He claims he first saw the light on the day that he shot a curlew in the Thames marshes. The bird had fluttered to the earth, still alive, after, in Iris Murdoch's memorable phrase, its 'pitiful surrender to gravity'. 'Curlew scream like children when they are wounded.' Fowles ran to the broken bird struggling in the mud and 'in too much haste I reversed my gun in order to snap the bird's head against the stock. The curlew flapped, the gun slipped, I grabbed for it. There was a violent explosion. And I was left staring down at a hole blasted in the mud not six inches from my left foot. The next day I sold my gun.'

It seems to have been this close shave with his shotgun and not the pain of the curlew that brought about his Damascene conversion. We have his word for it that John Fowles never again shot another bird or pinned another butterfly.[6]

With the enthusiasm of a convert he looked back 'in angry shame' on what he had once enjoyed. 'My own painfully slow progress from the sick days of my own little schoolboy *Wunderkammer* [Cabinet of Curiosities] showed me that nature is in fact not about collecting at all but about something infinitely more complex and difficult: being.' He now realised that 'these sacred relics . . . the pinned rows – obscenely like soldiers on parade – of butterflies and moths, were all lies.' He brought the same retrospective loathing to bear on his one-time passion for shooting and fishing, and came to see even the harmless twitching of birds as a pastime fit only for 'vain and narrow-minded fools'. In fact it was all far worse than mere foolishness. 'I came to nature through hunting it with gun and rod, and later by assembling mementoes of my "victories" over it. I don't think I realised that it was as if we were at war, just as the whole country had been with the Nazis . . .'[7]

It would be hard to find a greater contrast than between Fowles's agonising and the urbane assurance of that other great butterfly-loving novelist, Vladimir Nabokov. Nabokov collected butter-flies all his life from the time when he was a 'pretty boy in knickerbockers and a sailor cap' in Tsarist Russia to a 'hatless old man in shorts' in the Alps. Had it not been for the Revolution in 1917 he might have become a full-time entomologist attached to some Russian academic institution. In his adopted home of the United States he collected butterflies for Harvard and other American universities, and was, for a time, the official curator of Lepidoptera at Harvard's Museum of Comparative Zoology. In the process he discovered several new species, one of them (unfortunately since demoted to a subspecies) known as Nabokov's Blue – which he honoured with a short poem, 'On Discovering a Butterfly'. The great novelist's butterfly writings range from the poetic and sensual to terse descriptions written to scientific formulae.

Nabokov's novel, *Lolita*, was published only a few years before *The Collector*. Like that novel, it is full of metaphors of desire and capture; the latest Penguin edition of *The Annotated Lolita* actually has a fantasy butterfly on the jacket. The long car journeys of the novel were based on places where the author had collected butterflies in the 1940s. Nabokov includes references to butterflies that only a keen lepidopterist is likely to notice. Schmetterling, the German word for butterfly, is one of the false names used by his villain, Clare Quilty. Another character, Miss Phalen, borrows the French name for a moth. These subtle references, lurking in the verbal undergrowth, act as a kind of authorial watermark. They serve as a running metaphor for beauty and illicit passion.

There is, some would argue, something clinical and unsettling in Nabokov's fiction, especially *Lolita*. His mastery of language seems at times to be deployed with the crisp decision of a pin through a thorax. In his notes for the novel he measured the 12-year-old 'nymphet' like a specimen, so many centimetres' girth for the forearm, so much for the left leg. It was important to anatomise her to get the dimensions right.[8] The very word 'nymphet' has allusions to a butterfly. Many of the most beautiful European butterflies belong to the family Nymphalidae or 'the nymphs'.

In his autobiography, *Speak, Memory*, Nabokov's recollections of collecting butterflies in far-off Tsarist Russia have a peculiar intensity. As the son of an upper-class diplomat, he followed his father into exile after the Bolsheviks seized power. The memory of those childhood places and their bright butterflies, which he could never revisit, followed him in his subsequent wanderings like a faithful dog. Again and again in his work there is the same intense longing for a lost arcadia, a blissful, boggy place thronged with the blue, white and yellow butterflies of northern Russia. For Vladimir Nabokov collecting was not only a necessary part of the science of entomology but a link with the feelings for nature he remembered as a child; as part of the past which he wished to cling to, whatever else happened. Butterflies were the thread that brought the dissonant phases of his wandering life together and made them whole.

But Nabokov was also secure in regarding collecting as an indispensable aspect of exploratory science and so part of the search for truth and knowledge. For him butterflies were a touchstone to the intellect whereas, for Fowles, they seem to have been a source of shame. Fowles's idea that a respect for nature precludes any kind of 'use' is surely self-defeating, for we can hardly avoid, in his sense, 'using' nature all the time. To take his prohibition to its logical conclusion we would be 'defiled' even by picking blackberries from a hedge. The study of ecology would become virtually impossible. The lives of nature and humankind are entangled and intertwined – and that, arguably, is exactly why we care about it. In *What Has Nature Ever Done for Us?* Tony Juniper, a conservationist to the core, stood that argument on its head by suggesting that nature is valuable precisely *because* of its usefulness – though the unstated corollary, that species without uses are therefore without value, is not a comfortable thought.[9] Much of the scintillation which nature evokes in the weariest of spirits is surely found along that boundary where human experience meets the natural world.

John Fowles would not be the first writer to tailor his views to his audience and it may be that he felt he needed to show a radical face to the world for the booklet *Collector's Items*. At any rate when he came to review a book about butterfly collecting, *The Aurelian Legacy*, in which I had a hand, he was much more generous. 'I wish Nabokov had been alive to read it,' he wrote. 'I can almost hear him shouting for joy.'[10] (Reader, that was almost my proudest moment.) There, he was willing to pay generous tribute to the collectors of the past, admitting that without collecting there would have been, for example, no great theory of evolution, or at least, not then, and not in that way.

Given the butterfly collector's fall from grace, I think it is necessary to give him his due (in my experience, it is nearly always a 'he'). Before 1970, practically everything we knew about the lives of butterflies was obtained by studying, collecting and rearing them. The popularity of collecting was why we knew more about butterflies than any other group of insects. It was collecting, and earning enough from the sale of specimens to defray expenses, that drew Henry Walter Bates to the Amazon and led Alfred Russel Wallace to the islands of the East Indies and the eureka moment of how isolation produces new races and species. Insect collections helped Edward Poulton to discover the meaning of wing coloration and E. B. Ford and others to work out the processes of evolution in the field. Today the 'RCK' national collection is being studied with a new interest for the light it could shed on climate change. It is because there are specimens in collections that we know about the extinct British races of the Large Copper and the Large Blue, and the range of genetic variation in butterflies, and that the Glanville Fritillary once flew around London. It was the bankrolling of rich collectors by Lord Rothschild that led to the discovery of new butterflies all over the world.

Another collector, indeed another Rothschild, Nathaniel Charles, the brother of Lord Rothschild (of whom more in Chapter 6), ensured that we still have a few patches of natural fenland in East Anglia. A century ago Rothschild bought Wood Walton and Wicken Fen to preserve the habitat of certain rare moths from drainage and agricultural reclamation. He did so because these moths had a value – to collectors. Only later did these places become nature reserves and ultimately make possible the most ambitious re-wilding project in England today: the Great Fen Project. Without collectors and collecting nobody would have bothered preserving any fenland. Collecting gave a value to these rare insects and an incentive to conserve them, just as grouse shooting provides a strong incentive to preserve and maintain heather moors (without the income from grouse they would instead be over-grazed grassland covered in sheep). Collecting and conservation were not poles apart. You might even say that one was wedded to the other.

Lastly, chasing the gilded butterfly (and trapping those dusky moth) added substantially to the store of simple human happiness. It made naturalists for life out of children who might otherwise have lost interest. And without naturalists there would have been little incentive to preserve nature. *Pace* John Fowles, I do not think this is a shameful legacy.

COLLECTOR THE BOGEYMAN

All the same, you could say that Fowles has won. There are, unfortunately, psychopaths who kidnap women like the miserable protagonist of *The Collector* – and at least two of them claim they took that novel as their starting point.[11] But today, in Britain at least, the butterfly collector is as dead as his specimens. The net, along with the gun and plant-press, has become a metaphor for ecological sin. I would not want to bring back collecting. There are probably not enough butterflies left to sustain it. Those who manage nature reserves to improve the lot of butterflies would not take kindly to seeing the fruits of their labours vanishing into somebody's net and killing bottle – and rightly. Although butterfly collecting is not yet illegal, except where there are protective by-laws, it is no longer respectable. In crowded, ecologically diminished Britain, the days when we could collect butterflies and moths innocently, without feelings of guilt, are over. There is no point in trying to resuscitate the corpse of a bygone activity. We can gain the same kind of enjoyment from photography and from rearing insects in captivity.

Yet the shadow of the near non-existent collector still seems to haunt the public consciousness. I still meet volunteer wardens on nature reserves who tell me they are always on the lookout for the sneak butterfly thief. Sites on the Internet solemnly inform us that no one except an authorised leader of a group should be allowed to carry a net, and that even photographing protected species may be illegal since it counts as *disturbance* (though that might be hard to prove in law). We also experience legal obstacles to rearing the scarcer species since trade in their eggs or caterpillars has been banned under the CITES (Convention on International Trade in Endangered Species) rules. The countryside is not a comfortable place for the boy or girl entomologist today unless they confine their activities to *looking*.

It is against the largely imaginary threat of collecting that many British species of butterfly have been legally protected. Yet, far from helping the species, this form of protection simply distorts the nature of the problem – which is not collecting but habitat change or destruction. It is not fashionable to say so but every progressive farmer or forester or urban planner has done far more harm to butterflies, and indeed every other form of insect life, than collectors ever did. Butterfly collectors, like those who shoot or fish, had a strong interest in wanting to protect the natural habitat. No collector I knew behaved like Fred Clegg, ripping up butterflies for his own sadistic pleasure. They were without exception people who loved the natural world, though admittedly sometimes with the sort of intensity associated with collectors of any kind.

What is it that drives this unjustified fear of the bogeyman collector, from our legislators in Parliament to the voluntary warden on the nature reserve? There seems to be more to it than the desire to safeguard a species; no one gets so hot under the collar when it comes to collecting beetles, say, or shells. There seems to be a moral imperative to it, beyond any ecological rationale. Could it be something to do with the birthday-card image of butterflies as icons of beauty and freedom; as if catching them is morally on the same level as taking a potshot at an angel? Or might it be just another part of the over-protective nanny culture that tells us that nature must not be touched – which more or less guarantees that children grow up not knowing the names of flowers or trees or butterflies?

In Britain, whose butterflies have been very thoroughly studied and documented, collecting, arguably, has little left to offer to the pursuit of knowledge (though, as a result, there is much less interest in genetic varieties and races than there used to be). The obsession that once went into securing rows of meticulously mounted butterflies has been transferred to recording them. Local atlases routinely record butterflies in more and more detail, in some cases down to single-kilometre-square units, far outstripping any practical use that it may have in monitoring their numbers (it would not take long for any reasonably mobile butterfly to fly from one square into the next). Other obsessionists try to see every species of butterfly in a year, or, in the case of at least one person I met, *every* year. It is, to my mind, the same instinct to collect but transferred into non-lethal forms of activity.

Abroad, especially in the Third World, the law still sees the collector as the main problem. Butterfly collecting is now more or less banned over large parts of the world including places where it would still serve a useful purpose. In China, Indonesia and Brazil, for example, the collector is seen as a poacher or social pariah, and collecting without permission could involve you in a lengthy stay in some of the world's less comfortable prisons. India, whose remaining wildlife habitat is being eroded every day, has become super-zealous about safeguarding butter-flies and beetles from what it calls 'international poaching'. It is currently considering legislation that will effectively place practical entomology out of bounds even to its own citizens. Where, you wonder, will that leave the study and monitoring of India's declining butterflies?[12]

In such places, formal permission to collect is unlikely to be granted except to scientific expeditions and authorised research bodies and even then, things being what they are, the permit may not arrive in time, or indeed, at all. The countries that have devised the most complicated rules tend to be those least able to operate them efficiently. There is, you might notice, a rough equivalence between the zealousness with which countries guard their butterflies and their laxity in preserving the butterfly's natural habitat. It is tempting to assume that the real reason for banning collecting in such places has less to do with science and conservation and more with a memory of Empire and a sense that collecting represents Empire in another form: neo-colonials who seem to think they have a right to steal *our* butterflies, *our* nature.

Collecting is banned over large parts of Europe, too, although the rules vary from country to country, and sometimes even within a country. In Spain, permits are said to be a useful way of bringing in revenue, like a speed camera or a parking ticket. But in Croatia they say no, and mean no, unless you are part of a sanctioned research project. France, Italy and Scandinavia are more tolerant; you can take your net to those countries so long as you don't take any protected species and do all collecting outside nature reserves and national parks. Britain lies roughly in the middle; cynics might say we have solved the problem by eliminating the butterfly.

Butterflies have long been seen as symbols of freedom. But it seems to me that the metaphorical pin is in different hands now. The objective case against the collector is remarkably feeble. The best the prosecutor can do is to dredge up ancient tales of some long-dead maniac said to have collected hundreds of Large Blues or sold numberless chrysalids of Large Coppers. Yes, such species were heavily collected, and, no, collecting probably didn't have any lasting effect on their numbers. I doubt there is a single species of butterfly in the whole world that was over-collected to the point of extinction. Butterflies do disappear, and at an increasing rate, but that is because of people who are not collectors: loggers, developers, farmers that slash and burn and then move on, corrupt politicians, big corporations, people who care about nothing but money, or simply humanity swarming over the earth like ants. It was not collectors who drained the fens where the Large Copper once flew, or built low-cost homes on the dunes near the Golden Gate in San Francisco, the last haunt on earth for the Xerces Blue. Catching butterflies was never remotely as deadly as the crushing and obliteration of the places in which they lived. We might 'protect' them but we don't really protect them. Conservation thrives on figleaves. We surround a carefully selected list of butterflies and moths with the majesty of the law. And then we turn their habitat into a car park. The law creates butterfly gardens – but without any butterflies.

5.

Lady Glanville's Fritillary

My first butterfly book was the *The Observer's Book of Butterflies*. Although it was part of a children's series – the little books were cheap enough to buy with pocket money – the text was bone dry and as solemn as a sermon. Just occasionally the author made some remark that seemed to have strayed in from outside. Such was his disclosure that the last Large Copper in England had been 'taken' by a Mr Wagstaffe (what a rotter, you thought). Another stray remark mentioned a certain Lady Glanville whose will had been disputed on the grounds that she collected butterflies – which was seen as a clear sign of madness. 'Entomology was not much thought of at that time,' chuckled the *Observer's Book* man. 'Those who collected butterflies were apt to be regarded by their friends as – well, just a wee bit daft.'

She stuck in my mind, this entomological martyr who lent her name to one of our rarest and most beautiful butterflies, the Glanville Fritillary. Today, the chequered Glanville Fritillary is confined to the Isle of Wight where it flies on broken ground in sight of the sea in May and June. And if you miss it then, you can find its black, spiky caterpillars during the summer holidays, basking or feeding on their surprisingly common food plant, ribwort plantain. Many naturalists will know the butterfly, but the eponymous Lady Glanville is much more elusive. The best-known short account of her was written long after her death by Moses Harris in his famous book, *The Aurelian*. In eighteenth-century language, crowded with what we would regard now as misspellings, Harris presents her story as a kind of morality tale: of a noble woman of enquiring mind traduced by her ignorant relations whose reputation was rescued in the nick of time by Men of Science:

> 'This Fly [i.e. the Glanville Fritillary] took its Name from the ingenious Lady Glanvil [sic], whose Memory had like to have suffered for her Curiosity. Some Relations that was disappointed by her Will attempted to set aside by Acts of Lunacy, for they suggested that none but those who were deprived of their Senses would go in Pursuit of Butterflies. Her Relations and Legatees subpoenaed Dr Sloane and Mr Ray to support her Character. The last Gentleman went to Exeter, and on the Tryal satisfied the Judge and Jury of the Lady's laudable Inquiry into the wonderful Works of the Creation, and established her Will.'[1]

To this Harris could only add that Lady Glanville was also remembered as a gardener who specialised in irises, one of which was still known as 'Miss Glanvil's Flaming Iris'. On this telling, then, the story had a happy ending. It seems that John Ray had taken the fast coach all the way from Chelmsford to Exeter to save a lady from shame and at the same time convince the court that an interest in butterflies was not only sane but a praiseworthy contribution to scientific enquiry.

Unfortunately it isn't true. By the time Harris had got hold of it, the tale had become garbled. John Ray was in no position to intervene in a dispute over the will because by then he was dead; he died in 1705, four years before Lady Glanville's own death. The other person mentioned in Harris's story, Sir Hans Sloane (1660–1753), was certainly alive, and you could imagine him flourishing Ray's *Wisdom of God* as proof that a devotion to insects and natural history was a sign of learning and even piety, and not madness. But nowhere do Sloane's preserved papers mention a subpoena of the kind mentioned by Harris. Whatever the truth of the matter, it did not involve Ray, nor, in all probability, Sloane either.

By chance, two letters from Lady Glanville were found preserved among the papers of James Petiver in the Natural History Museum archive.[2] She had signed the longer of the two as 'E. Glanvile' and Petiver evidently believed that her first name was Elizabeth. In the 1960s, this set off a researcher, Bill Bristowe, on a wild goose chase to trace someone of that name in the parish archives. He failed to find any Elizabeth Glanville but did discover an Eleanor or Elinor Glanville living in the right place at the right time. She was an heiress of considerable property in the West Country but was not a titled 'Lady'. That confusion probably came about from the practice of eighteenth-century writers of beginning nouns with a capital letter; she was a 'lady', not a 'Lady'. She was born Elinor Goodricke in Yorkshire in 1654, had been twice married and was mother to eight children. During the latter part of her life she lived at Tickenham Court, a few miles inland from Clevedon, near

Bristol. The house still stands, an austere stone edifice with a medieval hall and a Tudor parlour, although the formal gardens where Mrs Glanville tended her irises are long gone. It was, and still is, a wonderful area for wildlife. Each day Elinor could walk down the hill to the flowery meadows of the Gordano Valley or ride through the coppiced ash-woods that line the overlooking ridge.

What survives of Elinor's letter – the writing has faded and two pages are lost – suggests that she wrote as she spoke, conversationally, rather breathlessly, and not bothering much with punctuation or paragraphs. The letter concerned a box of pinned insects which she was sending to Petiver, care of the Swan tavern in Cornhill (later, you may recall, to be the head-quarters of the Society of Aurelians). She was anxious about the carriage and wanted to give detailed instructions to ensure its safe arrival. She hoped to send a second box from Wales, then an unknown hunting ground, 'wch a friend has this somer got for me and promist to send'. She feared, modestly, that her bugs were unlikely to excite Petiver very much; it was all common stuff, she supposed. She mentioned her difficulty in keeping her specimens from the ravages of 'mites' and a 'whit crusty mould wch when I came to clean [them] broke al to peeces'. It is clear that the two enthusiasts were collaborating, Elinor by collecting insects from entomologically unknown parts of England and Wales, and Petiver by helping her to name them and featuring the new ones in his catalogues. Elinor Glanville comes across as deeply interested, slightly scatty and duly modest in her dealings with a Fellow of the Royal Society. In a second, shorter letter, she mentions a gentlewoman who was sailing to Virginia and who had promised to send her 'some fine Insects' from there. Like Petiver, she was interested in butterflies wherever they came from.[3]

By good fortune we also have a letter from Petiver's side. Evidently a friendship had grown up between the two and 'Madam Glanville', as Petiver calls her, had entrusted him with the care of her wayward son in London, possibly as his apprentice. This was proving no easy task. 'Madam we must make the best of a bad Markett,' warned Petiver. The boy had been disruptive. He had repudiated her friend's guardianship and disappeared. Petiver suggested having him 'cryed by the City Cryer' but was loath to undertake such drastic measures without her consent. He feared that the boy might have fallen into bad company and regretted that he had shown 'noe Obedience to yu'.[4] Was this the same prodigal son who later disputed his mother's will and tried to ruin her reputation?

In revealing these glimpses of the real Elinor Glanville, Bristowe also threw light on the famous disputed will. Elinor's marriage to Richard Glanville, her second husband, had been unhappy. Glanville seems to have been an unscrupulous and perhaps violent man; he once levelled a pistol at her and threatened to blow out her brains. He had, it seems, tried to lay his hands on her fortune by circulating stories of her supposedly eccentric activities and forcing their children to sign affadavits against their own mother. But Elinor had retaliated by turning her property over to trustees while retaining the right to direct her own affairs. In her will she bequeathed the estate to her cousin, Henry Goodricke, reserving only some small legacies for her children, and nothing at all for Richard Glanville. She died at home in 1709, aged about 55. The disputation case was brought by her eldest son Forrest but was heard not at Exeter (so Harris got that wrong too) but at the assize court in Wells three years later. Nor does it seem that the outcome was the happy one of entomological legend. The 'wrong' people won. Forrest had sought to prejudice the jury – all of whom, of course, were men – with witness accounts of his mother's supposedly disreputable behaviour. Nosy neighbours testified how she would dress 'like a gypsey' and wander out of doors 'without all necessary cloaths'. One described in detail how she would spread a sheet beneath the bushes and, with two apprentice girls, beat the boughs with a long pole 'and catch'd a parcel of worms'. By such means twelve good men of Wells were convinced that Elinor must indeed have been out of her wits. Forrest, who also asserted that his mad mother believed in fairies, and considered that he, her eldest son, had been turned into a particularly wicked one, thereby had his way. The will was set aside and the children came into their inheritance.

Elinor Glanville was staying in Lincoln sometime in the 1690s, perhaps on the way to visit her Goodricke relations in Yorkshire, when she caught the butterfly that now bears her name. Lincoln and the Isle of Wight are a long way apart and it seems as though the Glanville Fritillary was much more common and widespread 300 years ago – a reminder that there is nothing new in climate-induced changes of distribution. Elinor sent specimens of the butterfly to Petiver asking him to name it. It was new to him too and he called it the Lincoln Fritillary. Later on Petiver caught the same butterfly near Dulwich and promptly renamed it the 'Dullidge Fritillary'. Exactly when the name was changed to Glanville Fritillary is uncertain but it was already current by 1748 when an artist called James Dutfield made a beautiful print of the butterfly under that name together with its plantain food plant and its caterpillar.[5] It has been known as the Glanville Fritillary ever since: the only British butterfly to bear the name of a British naturalist.

The image of a lonely, eccentric woman beating the branches for caterpillars while her neighbours peeped from the bushes is tempting and amusing but it is probably misleading. We know so little about Elinor Glanville that we are free to make of her what we will. The writer Fiona Mountain certainly did in her 2009 novel, *Lady of the Butterflies*, 'a dramatic tale of passion, prejudice and death by poison, of riot and rebellion, science and superstition, madness and metamorphosis'. But the real Elinor was mistress of her own affairs, able to order her estate as she wished, and with the means to maintain it. For her, the pursuit of entomology may have been a form of emancipation. Nor was she really alone in her pursuits. She was cousin to Adam Buddle, the butterfly-collecting vicar of Fambridge, and was on terms of trust and affection with James Petiver who clearly saw nothing disreputable about a lady taking a keen interest in butterflies. He and John Ray both had the highest regard for her work. Although she published nothing that has survived she is known to have compiled one of the first local lists of insects, for the Bristol area, and her collection, according to Petiver, 'sham'd us all'.[6] A few of her butterflies are said to survive among the Sloane collection in the Natural History Museum.[7]

One more extract from Elinor Glanville's letters reminds us of how easy it must have been to be taken in by frauds or practical jokes in those early days. 'I rejoice to find by yr Catologue,' she wrote excitedly to Petiver, '[that] you have got Mr Charlton's blistered butterfly it being my perticular favorit.

I fear none of mine wil deserve to be put in yr tables. I wish I could procure you such curiositys.' What was this 'Charlton's blistered butterfly'? By chance it still exists and it is indeed a 'curiosity'. Petiver regarded this novel butterfly as a new species related to the Brimstone and described it as such. 'It exactly resembles our English Brimstone Butterfly,' he noted, 'were it not for those black spots and apparent blue Moons on the lower wings . . . it is the only one I have yet seen.'[8] It had been presented to him by 'my late worthy friend Mr William Charlton a little before his Death'. Even when no more spotted and mooned Brimstones were discovered, later authorities were sufficiently impressed with it to provide the butterfly with a formal scientific name: *Papilio ecclipsis* – the Moon Butterfly. They were all deceived. The mischievous Mr Charlton had carefully painted the spots on to the wings of a normal, male Brimstone. It took half a century for someone to spot the fakery. When the deception was pointed out, the deceived and angry curator, Dr Shaw, 'indignantly stamped the specimen to pieces'.[9] Fortunately someone had made copies of the Moon Butterfly and they survive to this day in the collection of the Linnaean Society where they are known as the Charlton Brimstones. And happily, Elinor Glanville's box of – as she supposed – worthless insects contained something much more interesting. They included the first-known specimen of the Green Hairstreak, the only British butterfly whose wings are truly green – and a lovely velvet-green at that. We could easily have had a Glanville Hairstreak too.

By the early 1700s, botany, and by extension, natural history, was already a respectable, even fashionable, pursuit for upper-class women. At the very apex of society, Mary Capel Somerset, the dowager Duchess of Beaufort (1630–1715), dedicated the long years of her widowhood to painting or sewing images of flowers and butterflies. She seems to have found a sense of peace in the natural world at the end of a long and turbulent life. She used her great wealth and influence to gather tender and exotic plants from all over the known world, from China, the East and West Indies, and North America, to fuss over in her hothouses at Badminton. She also spent long hours pressing and drying plants for her herbarium, hiring artists to paint the choicest specimens and herself preserving their likenesses in needlework. In similar fashion she commissioned butterfly paintings and lent her support to that struggling artist of insects and spiders, Eleazar Albin. And according to Richard Bradley, her friend and the founder of the University Botanic Garden at Cambridge, the Duchess also reared butterflies and moths in her glasshouses. She had, it was said, 'bred a greater variety of English Insects than was ever rightly observ'd by any one Person in Europe'.[10]

Badminton House and Elinor Glanville's Tickenham Court are only a day's ride apart. In country-house terms, the Duchess and 'Lady' Glanville were practically neighbours. They also had acquaintances in common, notably Petiver and Sir Hans Sloane who were both among the Duchess's suppliers of seeds and plants. We know about Mary Somerset's interest in butterflies from a few offhand remarks and from the survival of some of her paintings and needlework. Quite likely there were other wealthy widows in the West Country whom history has passed over, collecting and breeding butterflies, and stitching their likenesses on to tapestries. Titled women were prominent among the subscribers to Albin's *Natural History of English Insects*. Was there another centre of entomological excellence, not in the great institutions of Oxford and Cambridge, nor the taverns and coffee houses of London, but one based in the drawing rooms and glasshouses of England's West Country mansion houses?

THE DUCHESS AND THE PHILOSOPHER

The eighteenth century was the heyday of the curio cabinet, a time when anyone with sufficient leisure and resources could cram their homes with natural curiosities such as shells, fossils, stuffed animals and bird's eggs. As we have seen, most butterfly collections of the period were part of a wider arrangement of 'the productions of nature', sometimes including busts of the philosophers for extra gravitas. The greatest of all these collections belonged to Margaret Cavendish-Bentinck, Duchess of Portland (1715–85). She was the daughter and only surviving child of Edward Harley, second Earl of Oxford, a wealthy bibliophile, collector and patron of the arts. He had learned the foremost lesson of his age, which was to stay out of politics. Margaret's grandfather had shown where that could lead: impeachment and a cell in the Tower of London on a charge of treason.

Some say the collecting bug is inheritable; that there is such a thing as a 'collecting gene'. If so, Lady Margaret, the future Duchess of Portland, had it too. She was Daddy's girl. She accumulated yet more precious objects and manuscripts, but above all she loved nature. Her particular passion was for shells.[11] She was also a passionate botanist and gardener and, perhaps inevitably, she added insects, particularly butterflies and moths, to her quotidian interests. Few have been able to indulge their tastes as thoroughly as the Duchess of Portland. Like her father, she inherited one fortune and then, at the age of 19, married into another. Her husband died young and so, once the burdens of being a wife and mother were past, the Duchess was able to devote the twenty-three years of her emancipated widowhood to her 'beauties': her collections, paintings and books.

There are many formal portraits of the Duchess but they seem to present us not with one woman but several. In an early one, a miniature, she has a comely heart-shaped face with hair curled back from a broad, intellectual forehead; in another her face is long and haughty with those languid doe-eyes so admired at the time. In a third she poses in a simple green dress with a black feather in her hair. In the best-known portrait, painted when she was about 30, she is dressed to the nines in silks and jewels but seems impatient, as though looking forward to getting back to her salon. The plainest is perhaps the one as she saw herself, a marble bust in the antique manner with simply dressed hair parted over that high, intelligent brow as if she were a poetic Muse or one of the Graces, keeper of the temple of knowledge.

Margaret Bentinck shared the mindset of the day which was to catalogue the productions of nature as fully as possible. Her grand ambition was to create a palace of 'works of art and vertu', including at least one specimen of every animal, plant and insect known to exist. This Noah's Ark of objects filled the public rooms of her country seat at Bulstrode, near Gerrards Cross in Buckinghamshire. A jungle of living plants and ferns thronged the glasshouses and spilled over into the gardens outside, all ordered in the classical manner. There was an American Garden for New World exotic blooms as well as an aviary, a private zoo and a shell grotto she had built with her own hands. Her silver dessert service was inspired by entomology, the handles of her spoons and fruit knives creeping and crawling with gleaming silver-gilt bugs. By the time of her death in 1785, the Portland collection of natural objects, and art inspired by natural objects, was a many-splendoured thing, perhaps the finest ever made by a private individual at least until the advent of Lord Rothschild and his famous museum at Tring.

But almost immediately after her death, the Duchess's palace of wonders was broken up, auctioned off to pay the electioneering expenses of one son and the drinking and gambling debts of another. The sale took thirty-eight days to complete.[12] The house itself was sold a few years later, fell into neglect and was duly demolished. You might say the wonders of Bulstrode lived and died with her. Her shells and butterflies were sold on and scattered among lesser collections in Britain and Europe, never to be seen in public again, and, bit by bit, thrown away as so much junk. The only way collections can survive the decades is to stay intact.

To Margaret Bentinck, collecting was a means to an end. Her specimens were not so much stored objects as a library of conversation pieces. As her friend and fellow bluestocking Mary Delany remarked, the primary aim of her collection was to enlarge the mind: the house and all belonging to it was in effect 'a noble school for contemplation'. It was open on application to visitors, and they came in large numbers. The Duchess valued intellect above social position. Among those she persuaded to join her establishment as curators and in-house men of learning was the genial botanist Daniel Solander, who took over the plant collections and became drawing master to her children. She also recruited another notable botanist, John Lightfoot, author of the first flora of Scotland, as her chaplain and librarian. Thomas Yeats, a leading London collector, became her curator of insects while the artist William Lewin was recruited to paint her birds and their eggs, and perhaps also some of her butterflies for his *Papilios of Great Britain* – the first work devoted wholly to British butterflies. Her patronage extended through correspondence to a wider circle of botanists, naturalists and writers, including the Halifax naturalist James Bolton who sent her specimens of lichens and fungi. Bulstrode became known as the Hive, a metaphor for the Duchess's industry but equally appropriate to her passion for insects.

One of Margaret Bentinck's many correspondents was the famous Genevan philosopher and writer Jean-Jacques Rousseau. Rousseau was a keen botanist. The two of them collected plants together in the Peak District in 1766 and corresponded for several years, exchanging specimens and books, identifying plants and enjoying a mutual regard. Indeed on Rousseau's side it might have gone further than regard. 'I know one somewhat savage animal,' he purred, 'who would live with great pleasure in your menagerie, in awaiting the honour of being admitted one day as a mummy in your cabinet.'[13] His service as a 'herborizer' or plant collector was, he wrote, hers to command, while modestly conceding that her knowledge of plants was greater than his. Rousseau was the kind of person who could hardly pick a flower without turning it into philosophy. He saw closeness to nature as the touchstone for human happiness since, so he taught, the primitive state leads to greater gentleness and content – in short, everything a truly rational society should strive for. It was decadent civilisation that had led to wars and social injustice; the simple life was by comparison peaceful and virtuous. Hence, botany too was a virtuous occupation for it brought us face to face with nature, and nature, unlike human-kind, could never lie. To him it was obvious that country life was healthier than that of the city just as the simple-living Roman Republicans were more virtuous than Rome under the emperors, or, for that matter, the kings and emperors of his own time. Nature, he preached, was good for us because 'it detaches us from ourselves and elevates us to its author', the 'author' being, of course, the Creator. 'It is in this way, Madam, that natural history and botany have a use for Wisdom and Virtue,' he told the dowager Duchess of Portland.[14]

This pleasant correspondence, flattering to both parties, came to an abrupt end when the Duchess innocently sent the philosopher a book about propagating and growing exotic plants. She had supposed Rousseau to be as interested as she in new ways of propagating coffee beans, pineapples and rubber trees. But, as she was to find out, the philosopher's view on such things was above all moralistic. He objected strongly to the importing of exotic plants on the grounds that it 'deforms nature'. Just as man would never be free until he had shaken off the chains that enslaved him, so plants too ought to be 'domination-free'. Rousseau's idea of ideal gardening would be close to the present-day idea of a wild garden, informal and full of simple plants that grow locally or, at the very least, have not had their naturalness bred out of them. He returned the book.

Margaret Bentinck's title, if not her name, lives on in the famous Portland Vase, now one of the treasures of the British Museum, and also – and she would have appreciated the compliment just as much – in a small, pretty moth with greenish speckled forewings, the Portland Moth. The heyday of Bulstrode and its treasures is remembered, more tangentially, in a second, equally pretty moth, the Pease-Blossom, a single violet wing of which was found in a spider's web there (I once set unbelieving eyes on it, still preserved in the national collection). This shred of wing, for ever divorced from its body, became a long-running entomological joke. 'Where can I take the Pease-Blossom moth?' 'Look for it in the spider's webs at Bulstrode!'

KNOWING YOUR PLACE

When it came to gender relations – though he generously made an exception of the Duchess – Rousseau's ideas were no more enlightened than those of the age which he spent his life castigating. A woman's place, he insisted, was in the home, preferably in the boudoir or the kitchen. Women, believed Rousseau, have no capacity for abstract thought, and especially not in the sciences. Their natural sphere was in practical matters. He feared that unless women were domesticated, and constrained by modesty and a commensurate sense of shame, they would tyrannise men and make their husbands' lives a misery. 'Given the ease with which women arouse men's senses . . . men would finally be their victims,' he warned.[15] Mankind should shake off social constraints and return to a state of nature – but women should stay in the kitchen.

Eighteenth-century English society had fixed ideas about what was and what was not appropriate. A moderate scientific education – for the few women for whom it was available – was acceptable since the alternative, the Classics, was closed to them as an exclusively male preserve. Chemistry, for example, was regarded as a fit occupation for ladies partly because it was seen as analogous to cooking.[16] Similarly, botany was open to women as well as men, because cultivating and arranging flowers were among the approved pastimes – and also because botany, eighteenth-century style, was scarcely distinguishable from gardening. Flowers, it was conceded, lay well within a woman's capacities. The clergy allowed that nature study, too, was virtuous and open to women because, as John Ray had preached, it was conducive to religious feeling. Anyone drawn to flowers and butterflies must surely love the works of the Creator.

In this sense butterflies were seen as part of 'botany' and especially since rearing them involved 'gardening' in the potting and growing of food plants for the caterpillars. But even so there were limits. Rearing butterflies had appropriate connotations of childcare as well as gardening and so was an acceptable occupation. So were drawing, painting and embroidering butterflies. But killing them for a collection was not. One reason why shell collecting was so popular among enlightened women was that shells could be gathered without necessarily killing the animal inside. It was all right for a man to go out with his gun or his hounds and be praised for his manliness, but women were supposed to be gentler. The Duchess of Portland enjoyed searching for snails and shells by lakes and the seashore but there is no evidence that she chased butterflies. Her butterflies seem to have been acquired by purchase, gift or exchange.

A woman was not only limited to the 'gentler' aspects of natural history but there were also restraints on how knowledge could be acquired. Philosophical works, such as Ray's *History of Insects,* were unsuited to female minds. Instead, perhaps frustrated by their exclusion from science and philosophy, women set themselves the task of writing books attuned to their own sensibility and experience. Writings for women turned on family and conversation, of the sort that they might enjoy at home or among close friends. The Duchess of Portland was herself a member of a circle of intelligent women known as the Bluestockings who met for serious conversation in salons and drawing rooms, hoping thereby to 'establish women's intellectual independence in a socially acceptable form'. A Quaker writer, Priscilla Wakefield, specialised in this epistolary approach to science, seeing it as 'blending instruction with amusement'.[17] Hence, for example, she introduced the habits of bees by way of a dialogue between Clarice, a well-educated country woman, and her enlightened gentleman friend, Eugene. In this conversation Clarice is anxious that an interest in insects might be incompatible with her responsibilities as a woman and young mother. Eugene makes haste to reassure her that bees are an entirely appropriate subject for they, too, are good mothers, and do not their busy lives echo the economy of the household? The two of them agree to spend some time together after dinner, talking about bees, walking in the garden and observing the hives. Even so, warned Miss Wakefield, such knowledge was dangerous; ladies should use it with discretion. Parading your education in company was vulgar and a sign of ill-breeding. So keep it to yourself.

Given such constraints it is perhaps surprising to find that fully a quarter of the hundred or so persons listed as subscribers or 'encouragers' to Benjamin Wilkes's *English Moths and*

Butterflies, published in 1749, were women. And some of them did more than subsidise. Wilkes singled out a Mrs Walters who was renowned as a breeder of rare moths and, by implication, a supplier of source material for his pictures. It seems that, already by then, women had gained a reputation for their expertise in rearing butterflies and moths which surpassed the efforts of men. In this, as in much else to do with rearing, women held the advantage.

WOMEN WHO FOUND A DIFFERENT PLACE

As the story of Elinor Glanville demonstrates so well, women who interested themselves in butterflies and other insects could face social ostracism. In the resolutely masculine atmosphere of the Victorian age, they were excluded from participation in the new entomological clubs and forbidden to write papers for their journals. Hence women were unable to write up their discoveries under their own name. That is why we do not hear much about one of the best Irish entomologists, Mary Bell (1812–98): she was unable to publish her original discoveries about water bugs and dragonflies (she was the first to discover that water bugs could sing like grasshoppers). Her brother Richard got the credit instead.[18]

Fortunately, women could and did write books. Laetitia Jermyn (1788–1848), the self-styled 'Fair Aurelian', produced what became a standard text, *The Butterfly Collector's Vade Mecum* (or 'ready reference'). In 1848 Maria Catlow published a *Popular British Entomology* that was successful enough to be reprinted twice. Yet both authors felt it necessary to write in a style befitting a woman: which in Jermyn's case required long, tedious digressions into poetry and moral philosophy, and in Catlow's a prettified dumbing-down that limits its usefulness.[19]

As for rearing butterflies and moths, maternal solicitude reached its height in the quiet work of Emma Hutchinson (1820–1906), the only other person whose name is preserved in a British butterfly, in her case the golden midsummer form of the Comma known as var. (variety) *hutchinsoni*. Emma was the wife of the vicar of Grantsfield near Leominster in rural Herefordshire, and one of her few publications concerned what was, and what was not, proper entomological behaviour for a lady. Rather than collecting butterflies, like men, she suggested that women study their 'habits'. They should rear them through each stage of their life cycle and carefully note each change of skin and difference in form. It was not quite proper for a vicar's wife to procure eggs and caterpillars herself, but Emma could at least raise plants for those brought to her by various (male) helpers and keep them supplied with fresh food. She bred one small, drab moth, the Pinion-spotted Pug, through countless generations from 1866 to the end of her life, supplying all the collectors of the day with specimens. But the species she is associated with above all others is the Comma. The Comma butterfly is common enough today, but in the nineteenth century it was regarded as a rarity and at one time seemed to be in danger of dying out. The one place you could be confident of finding its spiky, brown-and-white caterpillar was on the leaves of the bines in Herefordshire's hop yards. She blamed the practice of burning the bines after the harvest for the growing scarcity of the butterfly. In perhaps the first attempt to conserve a British butterfly, Emma reared generations of Commas and sent boxes of caterpillars to correspondents in the hope that releasing the butterfly would safeguard the species. Hundreds more went 'to gladden other naturalists in their collections'.[20]

By breeding the Comma year on year Emma Hutchinson was the first to realise that it is double-brooded and that the two generations are different. The bright golden-orange form of the butterfly, easily mistaken for a fritillary, appears from caterpillars reared in the spring, and the darker form from those reared later in the year. Emma deduced, correctly, that it was day length that decided which form, dark or golden, would emerge; the golden one enjoyed the balmy days of high summer while the darker, less conspicuous generation is the one that needs to pass the winter undetected and emerge from sleep in the spring. Acknowledgement, of a sort, came when the Natural History Museum accepted her collection of butterflies while her notebooks and records, too, have been preserved in the library of the Woolhope Naturalists' Field Club.

A different kind of emancipation marked a trio of fearless, intrepid lady travellers and collectors: Margaret Fountaine (1862–1940), Evelyn Cheesman (1881–1969), and Cynthia Longfield (1896–1991). All three were of independent means and none of them ever married, so they were free to do more or less as they pleased, and to travel where and when they wished. They were also lucky enough to live through a time when the going was good: a golden age of exploration and discovery when steamships, railways and modern roads were opening up distant parts of the world for the first time. It was the time when Lord Rothschild and other wealthy patrons were financing collecting expeditions to remote places and gorgeous new species of butterfly were being described almost daily. For these women, travel offered fulfilment and freedom from the suffocating constraints of home.

Evelyn Cheesman wanted to be a vet but was unable to train at the Royal Veterinary College because it did not admit women. Instead she got a job at London Zoo's insect house – though even that meant bending the rules; she was the first-ever woman on the staff. Thereafter she embarked on a series of solo expeditions to Papua New Guinea, the New Hebrides and other tropical islands of the Pacific, braving leeches, tarantulas and malaria to collect insects for the Natural History Museum. A small, skinny woman, generally dressed in a mackintosh and sackcloth trousers, she was not interested in home comforts. Encounters with alleged cannibals and incestuous tribes merely aroused her scientific curiosity. Like her contemporary, Freya Stark, she defrayed her expenses partly by writing travel books. By the end of her travelling career she had donated 70,000 insect specimens, many new to science, to the museum. The least they could do was to name several of them after her.[21]

Cynthia Longfield – aka 'Madam Dragonfly' – was from a landed Irish family with a great house at Castlemary in County Cork. Perhaps *her* emancipatory moment came when Irish revolutionaries burned the house down. She too decided to venture riskily into some of the wilder corners of the world that looked interesting for insects: the South Pacific, Brazil's Mato Grosso, and the interior of Australia, not to mention a six-month solo expedition to Uganda. Her great passion, the abiding love of her life, was for dragonflies. Hitherto insect collectors had never bothered much with dragonflies because, unlike butterflies or beetles, their colours fade after death. Like her friend Evelyn Cheesman, Cynthia Longfield collected for the Natural History Museum, packing each delicate object between wadding in airtight boxes and sending them off by bearer before the heat and damp destroyed them. Unlike Cheesman, she had no great talent for writing and accounts of her exploits are largely confined to professional journals. But her popular book on British and Irish dragonflies helped to nurture interest in living dragonflies and inspired other, more ecologically minded pioneers such as Norman Moore and Philip Corbet, to specialise in dragonflies. Today, like butterflies, these lovely insects are firmly in the spotlight and have their own society and recording scheme. They have become, par excellence, the birdwatcher's insects, the ones you watch with binoculars. Cynthia Longfield lived to see the formation of the British Dragonfly Society in 1983, living another eight years to reach the grand old age of 96.[22]

The best-known of the three is Margaret Fountaine although most of her fame is posthumous. 'To the reader – maybe yet unborn – I leave this record of the wild and fearless life of one of the "South Acre Children" who never grew up, and who enjoyed greatly and suffered much,' she wrote at the head of the journal of her travels which she kept for sixty years (South Acre, near Norwich, was her family home).[23] After her sudden death while catching butterflies in Trinidad she left her large collection to the Norwich Castle Museum along with a locked japanned box which, on her instructions, was not to be opened until forty years later. When the time came and it was duly opened and inspected, the box was found to contain her journal, including a full account of her travels with her faithful dragoman, the Syrian Khalid Neimy, her 'dear and constant friend'. It was their intimacy, regarded as shocking at the time, which was probably the reason why Fountaine had embargoed the journal. Highlights from it were turned into a best-selling book, *Love Among the Butterflies*, although anyone hoping for a tale of illicit passion in exotic places must have been disappointed. The event that impelled Fountaine's wanderings was a broken heart; she had been rejected by a man she hardly knew but had decided he was the 'One and Only'. Her relationship with the already-married Khalid Neimy was apparently more like brother and sister.

When Norman Riley, the museum's curator of insects, met her in 1913 he expected to find what he ungallantly described as 'a well-worn battleaxe'. Instead he found a tall, pale, shy, attractive woman with an air of frailty and melancholy. Slightly more self-conscious than Evelyn Cheesman (but only slightly) she chased butterflies in a man's checked cotton shirt with a long, striped cotton skirt, both with extra pockets sewn on, topped off with a pair of cotton mittens. She sustained herself on her travels with nips of brandy from her flask. A graceful retirement was never what she had in mind and she got her wish. She died with her butterfly net in her hand, aged nearly 80, one hot day on a mountainside in Trinidad.

6.

At the Sign of the Chequered Skipper

The last Chequered Skipper in England flies on the green of a small village in Northamptonshire. Strictly speaking it doesn't actually fly; it swings from about eight feet up. This rugged skipper is made of old hobnails which reproduce the texture and colour of the small brown butterfly with surprising fidelity. It swings above the doorway of the Chequered Skipper pub in Ashton, near Oundle. There are surprisingly few pubs named after butterflies – bees and their hives are the popular pub insects – and this one wasn't always called the Chequered Skipper. Its original name was the Three Horseshoes. Stone-built and thatched, according to the specifications of the estate owner, it stands at the edge of a landscape that was once fantastically rich in butterflies. The nearby woods contained six kinds of fritillary and all five British hairstreaks as well as rarities such as Wood White and Duke of Burgundy – and, of course, the real live Chequered Skipper. At one time, the now extinct Large Blue flourished at nearby Barnwell, and Large Coppers and Swallowtails flew over the fens not far away. Most of them are gone now, victims of butterfly-unfriendly practices that have boosted crop production at the price of sterility. The lone butterfly on the pub sign is their monument. In 1996 the old pub burned down but it was rebuilt by a 'heritage' architect in replica form, more or less, and the new owners were happy to retain the old name. So the iron butterfly still swings above the village green to remind us of past glories.

Chequered Skippers were always harder to find than pub signs. Even in the best of times this was a scarce butterfly that fluttered and perched and then vanished in a twinkling. The prettiest of the skippers, in its deft chequerwork of light and dark brown, it was confined to the corner of eastern England bounded by Lincoln, Northampton and Cambridge. Its first captor was Charles Abbot, vicar of Goldington, who caught it at Clapham Park Wood near Bedford one sunny day in May 1797 during his 'first season as an Aurelian'.[1] He claimed discoverer's rights to call it the Duke of York Fritillary to match the Duke of Burgundy Fritillary, for it shared that butterfly's stubby build and basic colours. But he was overruled. The then presiding genius of British Lepidoptera, Adrian Haworth, was not minded to call the new species a fritillary when he knew it was a skipper, and so he insisted that it be called the Chequered Skipper. Nineteenth-century collectors found the new skipper in many woods within its limited range, especially those with an abundance of its favourite flower, bugle. It seems to have thrived best in woods that were regularly cut over as coppice and with a network of broad rides for the timber carts. But after 1914 the practices that had sustained these open, flowery woods began to fall into abeyance. The trees grew up and shaded the ground, or were felled and replanted with even shadier conifers. The butterfly became rarer and rarer. It even died out on nature reserves. By 1976 the Chequered Skipper was finally pronounced extinct in England, leaving butterfly watchers scratching their heads at the rapidity of its demise. Perhaps the great droughts of summers '75 and '76 were the *coup de grâce*. And that might have been that – another butterfly gone – except that, to everyone's surprise, the Chequered Skipper had also been discovered in Scotland, in 1942. The extinction of the English race revived interest in the little-known Scottish Chequered Skipper. Butterfly surveyors discovered that it was more widespread than they knew, though confined to the mild, rainy glens of Argyll south of Fort William. It is probably Britain's least familiar butterfly. For most of us, Fort William is a long way away and the weather often uncertain when you get there.

The man who bought the Ashton estate and built the Chequered Skipper pub was Nathaniel Charles Rothschild (1877–1923). Like so many of his family, he was an able and versatile businessman with self-discipline and plentiful good sense. In due course he succeeded his father as head of N. M. Rothschild & Sons. But, like several other members of his immediate family, Charles Rothschild was born with the collecting bug. In his youth and early manhood he collected beetles, then butterflies and moths, and finally, and with a still greater zeal, he collected fleas. He was an authority on the butterflies of Hungary and Transylvania, where he met his future wife. He was the first to rear and note the life stages of several East European species, including the lovely Pallas's Fritillary. And once he had turned his mind to the challenging world of fleas, he discovered the most important flea of all, the Oriental Rat Flea, *Xenopsylla cheopis*, probably the species that had carried the bubonic plague bacillus and wiped out a third of the population of Europe. It was during his travels around Britain in search of beetles, butterflies and fleas that Rothschild became aware, to a degree few realised then, that our wild places were not going to last for ever. Suburbs were spreading around all the major towns and farming improvements were draining and reclaiming land that would once have been impossible to cultivate. Rothschild realised that we needed to know more about the best places for wildlife both in Britain and indeed (for such was the breadth of his vision) throughout the then vast British Empire. He was determined to do something to try and save these habitats and in the process he founded what would one day become the Wildlife Trusts. You might say that he invented nature conservation.[2]

Charles Rothschild was in his 20s when he discovered Ashton Wold. The story goes that, enchanted by the local woods and the abundance of rare butterflies he found there, he asked at the inn who owned the land. He was told that it belonged to a very rich and secretive family who seldom sold property because it never needed to. He must have been cheered when he discovered that the family in question was in fact his own. His grandfather Lionel had, unknown to him, in 1860 acquired the estate.[3] Charles set about building a family home there. For the estate workers, he engaged a reputable architect to design cottages of stone and thatch with running water and a degree of comfort, well in advance of the norm at that time. He built the pub on the green which doubled as a village stores and post office. A mile away, on a south-facing rise, he built a mock-Elizabethan manor house called Ashton Wold. Three large ground-floor reception rooms faced a sunny terrace and beyond it a panorama of unspoiled England: meadows, woods and water. Ashton Wold was a secluded but comfortable house, reachable only by a narrow, potholed track (which stayed potholed and unsurfaced throughout the twentieth century) and visible only on the closest approach. His daughter Miriam, the famous entomologist, was born there in 1908 and spent most of her childhood summer holidays there too.

Because his older brother, Walter, Lord Rothschild, was constitutionally unable to conduct business, or at least not to the standard their father expected, Charles, the younger son, the 'willing, conscientious, oversensitive Charles', had to shoulder the burden of their banking empire. But he was too highly strung to thrive in the world of international finance and he worked too hard. By his mid-30s he was in the throes of a depressive illness which was diagnosed as a form of encephalitis, an inflammation of the brain that causes headaches, fever and fatigue, as well as, in his daughter's words, 'prolonged bouts of obsessional thinking'. He had also begun to suffer from the then little-understood condition of schizophrenia. One day, disconsolate and unable to bear any more, he went into his bathroom at Ashton Wold, closed the door, picked up a razor and cut his throat.[4]

His daughter Miriam was 15 when it happened. She shared her father's love of butterflies and also, if a consuming passion can be said to be love, of fleas. 'For two years after his death I completely gave up natural history,' she recalled. 'I thought it was a cruel and terrible thing to catch all these wonderful butterflies and stick pins through them.'[5] And so instead of collecting butterflies like her father and uncle, Miriam watched them. She might not have become a scientist at all – literature had been her initial preference – but for her brother Victor, who decided to involve her in his school holiday task: to dissect a frog. 'We killed this luckless frog by chloroforming it,' she recalled. 'We made the dissection and I was so thrilled with what I found – the blood system which you can see without any trouble because it was so near the surface of the inside skin of the frog – I went straight back into zoology with a pair of scissors in my hand.'[6]

In her scientific work, which continued over a long lifetime, Miriam Rothschild chose butterflies both because they are ideal insects for study and because, her dogs apart, they were her favourite forms of life. She took up her father's interests and causes and added some of her own. 'All my life,' she recalled, 'I have tilted against hopeless windmills.' She campaigned with her Oxford colleague and friend, Professor E. B. Ford, for the legalisation of homosexuality. She regarded same-sex love as 'quite natural, for it occurs throughout the animal kingdom'.[7] She thought the way we treat farm animals, especially chickens, was 'terrible, inexcusable' and did her best to oppose the more barbaric practices in research laboratories, factory farms and slaughterhouses. She gave up eating meat and refused even to wear leather, hence the short white wellington boots that, along with purple Liberty silks and headscarves (the apparel of the Purple Emperor), became her sartorial trademark. Another of her causes was the treatment and care of mental illness. In 1962 she founded the Schizophrenia Research Fund, an independent charity dedicated to advancing the better understanding and treatment of mental illness, and schizophrenia in particular;

after her death in 2005 it was renamed the Miriam Rothschild
Schizophrenia Research Fund.

I met her for the first time one cold January day in 1994,
having bumped and bounced down the rutted and frosted
track to Ashton Wold. Miriam wanted to pick my brains
about something and had invited me to one of her famous
lunch parties. She was wheelchair-bound by then, having
broken her hip. It was my first experience of the eccentric
lady in the lonely, now overgrown house, though I knew
about some aspects of her work. Many people found her
combination of dissent, directness and sharp intelligence
disconcerting and even scary. But I never found her anything
other than kind, fascinating and, in her way, generous. We
met in a comfortable downstairs room that doubled as a
library, with piles of new books heaped on the long table. A
large fire roared under the mock-Elizabethan chimney.
Conversation wandered from the best way to milk a cow to
the curious habits of flukes; from the crookedness of museum
curators to the sly egotism of scientists; before turning with
characteristic abruptness to the extraordinary intelligence of
Miriam's miniature collie dogs. Has *your* dog ever made a
plan? she asked me, earnestly. She took it for granted that I
had a dog. She said of her father that he had the knack of
making people feel important and *clever*. Miriam had the same
gift. She was intent, exact, imperious, humorous (again, in
her way), gruff, decided, modest and sometimes surprisingly
poetic. She had a dry chuckle and sometimes told 'roguish'
stories. But she was not entirely attuned to the world beyond
Ashton Wold. For instance, she disliked using the telephone,
'this instrument', as she called it. When she reached what

she judged was the end of a phone conversation she abruptly hung up. You got used to it.

Perhaps you can catch an echo of what she was like in her list, made for a radio programme, of the world's greatest wonders. They were, in no particular order, the Jungfrau; the jump of a flea; the life cycle of the parasitic worm *Halipegus*; carotenoid pigments; the tigermoth ear mite; Jerusalem glimpsed through a sandstorm; and the Monarch butterfly. There had to be at least one butterfly. 'I must say,' she told me, 'I find everything so interesting. I've never been bored. Or tired.' Actually she sounded tired as she said it. She also claimed that the secret of happiness lies in knowing *how* to be bored, which might be another way of putting it.

Miriam was one of the best-known scientists in Britain. Yet she regarded herself as an amateur. She was largely self-taught. Her father had suffered at school and held a low opinion of schoolmasters and examinations. Like so many women of her era and class she was educated at home by governesses. She did not even have a formal degree; all eight of her doctorates were honorary. Because she found it impossible to make up her mind, she had enrolled for two opposing university courses at Oxford, in zoology and English literature, but failed to offer herself for final examinations in either. As she explained, 'You always wanted to hear somebody talk about Ruskin when it was time to dissect a sea urchin.'[8]

Miriam's lifelong fascination with invertebrate parasites began inauspiciously at the Marine Biological Station in Plymouth where she measured snails and investigated the flukes that infest their innards. At the same time she was engaged on a monster project to catalogue her father's collection of fleas, work which required close study of insect anatomy. For years she stared down the microscope at 'the backside of fleas'. She kept the live fleas in bags in her bedroom. She claimed to find beauty where few of us would even want to look: in the internal anatomy of flukes and parasitic worms (she compared the guts of some hideous worm to 'a chamber of glittering candles') or the strangely entrancing organs of fleas. The baroque complexity of a flea's penis might well have been another of her world wonders – though, perhaps to balance things out, for the cover of her book, *Atlas of Insect Tissue*, she chose a close-up of a flea's vagina.

Miriam, I think, had one of those rare, truly original minds that a conventional education might have crippled. As an autodidact she managed to retain her ardent sense of enquiry allied to an unusual breadth of interests. But there was something else there too, something rarer and harder to define, a powerful synthesising imagination, allied to a poetic, almost soulful sensitivity. Perhaps that is why, unlike most scientists, she could write like an angel in a clear, expressive, exact style that she claimed to have learned from her favourite author, Marcel Proust. Her personality shines from every page in her lively book about parasites, *Fleas, Flukes and Cuckoos*, which she dedicated to her dead father. When the nation made her a Dame, Miriam accepted the honour with characteristic insouciance. She sent me a card of a flea with little fleas on its back. I think I might have been one of those little fleas, and if so I was proud to be there.

One of her father's legacies was a plan to document wildlife sites 'worthy of protection' – the 'good spots', as he had referred to them. He hoped to get the recently established National Trust interested in acquiring some of the properties for the nation. When the trust showed itself completely uninterested in doing that, he set up a special Society for the Promotion of Nature Reserves and kick-started the process by buying certain tracts of wild land for preservation out of his own pocket. Many of his 'good spots' were good for one thing above all else: butterflies and moths. They included Wood Walton Fen, where Rothschild hoped to reintroduce Large Coppers brought in from Holland, and Meathop Moss in the Lake District, which was noted for a special variety of the Silver-studded Blue in which the females, instead of the usual brown, were an electric blue. Good spots closer to home included the places where Charles had collected Chequered Skippers. But Ashton Wold was not among them, perhaps because, as well as wishing it to remain private, he saw his broad acres as relatively ordinary countryside. If some of these places had been made nature reserves then, rather than half a century later, the conservation of butterflies would have been given a sizeable head start over other wildlife. But instead Britain went to war in 1914 and the project was forgotten. The demands of two world wars put nature conservation back by half a century.

Over another lunch at Ashton Wold, this time tête-à-tête, Miriam sought my assistance as a supposed 'expert on nature reserves' to relocate the lost 'Rothschild reserves' and to help her write a book about them. She hoped that something might be done about restoring some of them, perhaps by means of tax incentives for their owners. I agreed to work as her research assistant for a while. That done, she turned to other things on her mind: the role of pyrazines in helping us remember smells; the problems of poachers on the estate and the uselessness of the local police; and, finally, the recent discovery of cyanide molecules in outer space. 'It gave me a good ending for my burnet moth paper,' she said.

Many of Charles Rothschild's good spots had gone; some were now ploughed fields or suburbs; others *had* eventually become nature reserves as he had hoped, but minus many of the butterflies he had wanted to protect. The glory that was wild England in 1900 had lost some of its shine a mere hundred years later. Our joint book, *Rothschild's Reserves*, which Miriam subtitled 'Time and fragile nature', tells the story.[9]

This had seemed to me to be the perfect opportunity for Miriam to write an intimate biography of her father, just as she had done for her Uncle Walter in *Dear Lord Rothschild*.[10] But, to my surprise, she started and then stopped. 'I've had a terrible rethink,' she announced in one of those sticky phone conversations that always ended with a clunk and the dialling tone. 'It's all too long ago. No one will be interested now.' She proposed instead to pen just a few pages on the founder of our Grand Design and leave it at that. I did my best to dissuade her. There would be a great deal of interest, I protested, in the rather mysterious figure of Charles Rothschild, the effective founder of nature conservation in Britain. His 'lost reserves' were one of the great might-have-beens of rural history. If only he had lived, if only there hadn't been two world wars to fight, if only the National Trust had been as interested in wildlife as it was in stately homes. No, insisted Miriam, no one cares about that now. So our book became rather like *Hamlet* without the prince: plenty on the design but disappointingly little about the man behind it.

There were other reasons why she never wrote that

biography of her beloved father. Miriam was not well. She was gradually losing her sight and did not expect to live much longer. Moreover the felicity and grace in her writing which had once seemed so effortless were slowly deserting her. She plodded, then changed her mind, plodded a bit more, then changed it again in a different direction. Perhaps it was too painful a subject still. She would never write her own biography either, she said. There were too many skeletons rattling in the cupboards of Tring and Ashton Wold: her father's suicide, her uncle's blackmailers, the destruction of her mother's family at the hands of the Nazis, the last of which had destroyed her faith in the goodness of God. Ashton Wold had (she said) seen no fewer than three mysterious deaths. There was a ghost that had appeared from the sink and grabbed her hand. The house was now creepy in more ways than one. Much of it had literally vanished from view behind a leafy shroud of creeper. The disappearing house is, in its way, a fit metaphor for the fading memory of Charles Rothschild. Since most of his personal papers are now lost, and all those who knew him are dead, his story will probably never be written (though I think someone, some day, will write Miriam's).

Rothschild's Reserves was launched in October 1997 at a party in Sotheran's bookshop in Piccadilly with the whole Rothschild clan, young and old, in attendance. But that was not quite the end of it. Miriam had got the Royal Society to support our idea of choosing a dozen or so of these 'lost reserves' for restoration. We hoped that it might become an appropriate project for the Millennium, a nod (as it were) from one end of the century to the other. Sir John Krebs, then head of NERC (the Natural Environment Research Council), seemed interested, and even nominated someone to head the project should we manage to attract Lottery funding for what, by his calculation, could be around £40 million. That someone was none other than Jeremy Thomas, the man who reintroduced the Large Blue to Britain and who has done more than anyone to reveal the intimate lives of British butterflies. But we failed to get the funding we needed. Later, Jeremy told me what had happened. Our own grand design had, it seems, gone all the way to the Cabinet. The Secretary of State had apparently backed the plan but these were the nervous last days of the Major government, pilloried in the papers and unpopular in the country at large. The name of Rothschild, which had been such an opener of doors in 1912, now smacked of privilege. How would it look if the media heard that £40 million of Lottery money was to be spent on an 'elitist' Rothschild project? And so Miriam's resurrected reserves became another 'might-have-been'. When the Wildlife Trusts celebrated their centenary in 2012, proper homage was paid to the far-sighted man who began the process that would eventually flower a full half-century after his death. Appropriately, Tim Sands's book, *Wildlife in Trust*, opens with the portrait of Nathaniel Charles Rothschild.[11]

MIRIAM'S BUTTERFLIES

In her entry for *Who's Who*, Miriam gave as her 'hobby', 'Watching butterflies'. As a child she had a dream in which flowers magically broke away from their stalks, turning into butterflies and flying up into the sky. They were her 'dream flowers'. When, as a grown-up scientist, Miriam watched butterflies it was always with intent. She wanted to know what their colours were made of and what they were for. She wanted to know what lay behind the subtle scents of some butterflies and how they communicated with one another. Her chosen method of enquiry was biochemistry. Because such work demanded advanced equipment, and Miriam generally worked from home, she drew collaborators into her web, sometimes from Oxford and other English universities, sometimes from abroad. Miriam was fascinated by the hidden chemical arsenal that butterflies and moths can call on. If only we could detect their trails of scent, she wrote, we would 'hear' 'hundreds of [moths] calling frantically to one another directly darkness falls'.[12] If you squeeze a ladybird very gently the aroma will linger on your fingers for days. Some butterflies can play the same trick. A male Green-veined White releases a powerful, lingering whiff of 'lemon verbena'. The Peacock emits a musky scent that makes some people think of chocolate. Nabokov wrote of the 'subtle perfume which varies with the species – vanilla, or lemon, or musk, or a musty, sweetish odour difficult to define'.[13] Butterflies seem to communicate through such sprays and trails of scent, their odours defined in the memory by those mysterious chemicals, the pyrazines. Some moths can detect chemical signatures from hundreds of yards away. The early entomologists referred to such invisible trails as vapours (one species, a champion of this mothy art is actually called the Vapourer) and used a virgin female in a muslin cage to attract males from all around. There is a world out there that we poor humans can hardly glimpse with our senses. 'We stand awed and envious,' wrote Miriam, 'trying to catch more than a few clicks and ripples of sound with our inferior ears, and sniffing ineffectually on the perimeter of the virtually unknown world of chemical messages.'[14] She liked to imagine an 'umbrella of scents' that must hang over a meadow on a sunny summer's day, some attracting, some warning, each an encoded means of communication that butter-flies pick up with their knobbed antennae and their sensitive feet, and instinctively understand.

Previous generations of entomologists had managed to crack some of the colour codes of butterflies: the ones that say 'beware' or 'don't touch' or alternatively serve as camouflage or to mimic another, more dangerous insect, such as a bee or a hornet. They knew that some species laid their eggs on poisonous plants and that the toxins could be passed on by the caterpillar to the adult butterfly. You could even test whether a species was distasteful by cautiously sampling its bodily fluids. As far as I know Miriam never went to the lengths of Professor E. B. Ford in chewing various butterflies and moths to find out which were genuinely nasty-tasting and which were just bluffing. But she did demonstrate how widely these chemicals occur among butterflies. Many British species are aposematic, that is, they are conspicuous in order to advertise the fact that they are unpalatable. They advertise their toxicity with bright warning colours. The brilliant orbs of the male Orange-tip butterfly, for instance, not only resemble our no-entry road signs but have, essentially, the same purpose. This butterfly contains smarting mustard oils passed on from hedgerow plants by its caterpillar. The Large White's equally nasty taste comes from the bitter oils of cabbage and other brassicas. Both butterfly and caterpillar will burn the tongue of any bird that pecks it. Chemicals seem to have a second important function: they inhibit bacterial or fungicidal infection, like built-in antibiotics. Miriam discovered that burnet moths, which are brimful of cyanide stored from the vetches they ate as caterpillars, are resistant even to viral disease.

The most toxic butterflies and moths often contain a red pigment. Miriam wondered whether the colour red is learned more rapidly than other colours, especially when it is associated with an unpleasant experience. To her nose at least, the only moths that leave an unmistakable whiff behind are the tiger-moths, especially the Garden Tiger-moth, which has bright red hindwings as well as a defensive rattle that even our own ears can pick up. Tiger-moths can survive a lot of mistreatment. Miriam wrote of one individual which had been caught, bitten and then spat out by a bat and yet recovered long enough to lay its quota of eggs before expiring. Even more tough and toxic is the Monarch butterfly, famous for its long-distance flights across North America, whose caterpillar feeds on milk-weed, among the world's most poisonous plants. Miriam, working as usual with a university biochemist, was able to prove that the Monarch sequesters and stores heart-stopping poisons (cardiac glycosides) from the plant as part of its armoury. You've got to be very hungry to want to eat a Monarch.[15]

To assist her studies of butterfly colours and scents Miriam bred butterflies in the glasshouses at Ashton Wold including industrial numbers of Large Whites, Monarchs and the pretty, wide-winged American butterflies known as Heliconids. But she lamented the lack of wild butterflies she remembered from child-hood. Like so much of eastern England, the unspoilt places of Ashton Wold had suffered during the war when the old meadows were ploughed and much of the natural woodland was felled for timber. Miriam planted the gardens with flowers attractive to butterflies and moths such as dame's violet, red valerian and tobacco plants. But there was now little to attract them in the fields beyond the house: 'Not a flower in sight. Modern agriculture had bulldozed, weed-killed and drained all the flowers out of the fields that I'd known as a child. We were living on a snooker table.'[16]

Fortunately one area, known as 'The Roughs', had been spared from weedkiller and so could form a reservoir of seed for the missing flowers. Miriam's solution, which was novel then, was to harvest wild-flower seeds from such places and sow them into prepared plots in the barren, flowerless fields. She became, in her words, a 'grass gardener', a creator of meadows where 'the tides of grass break into foam of flowers'. It worked; the idea inspired others, among them Prince Charles who adapted some of her ideas in his garden at Highgrove. After Miriam's meadow gardens appeared in the Chelsea Flower Show, the idea became positively fashionable. Local authorities now routinely use her 'Farmer's Nightmare' seed mix for sowing in public open spaces. Private landowners and farmers, too, use wild-flower seed, creating new meadows that have become refuges for butterflies such as Meadow Browns and the now sadly misnamed Common Blue.

Among the butterflies that returned to Ashton Wold was one of Miriam's favourites, the Marbled White. She had long suspected that this black-and-white chequered butterfly of chalk and limestone hillsides is aposematic; that its brightness must have some warning purpose. Her conjecture was strengthened when one of her tame birds rejected both the butterfly and its chrysalis (though mice, she noted, guzzled both without hesitation). But the Marbled White's caterpillar feeds on grass, the same kinds of grass that are grazed by cattle and sheep. If the butterfly is toxic, then how and from where does it acquire its toxins? Miriam found the answer in almost the last piece of research she ever undertook. The toxic ingredient in the Marbled

White was finally detected, isolated and tested, and found to be a chemical called ioline. This is produced in nature not by grass but by a fungus, *Acremonium*, a mildew that grows on grasses, and especially on red fescue, the tufted, narrow-leafed grass favoured by the Marbled White caterpillar. It seems that the butterfly can detect this fungus and choose grass with the right 'scent' on which to lay. The confidence with which it shows itself to the world allows the butterfly to fly more freely than the other, more subdued, members of its family. I remember the delight with which Miriam told me the Marbled White was back at Ashton again, and her still greater delight in at last discovering its secret. I hope it will not sound over-sentimental if I suggest that this last discovery came as a gift. It was if the natural world was rewarding her for recreating the flower fields of Ashton that once again dance with butterflies.

UNCLE WALTER'S BUTTERFLIES

Perhaps the greatest honour a naturalist receives is when his name is attached to a newly discovered species. Miriam's uncle, Walter, Lord Rothschild, was so honoured many times. His tally includes a subspecies of giraffe, a bird of paradise, a scarlet climbing lily and a gorgeous slipper orchid, as well as – less flattering perhaps – a white intestinal worm. His name was also conferred on a butterfly and a moth, both appropriately giant-sized and colourful. The broad wings of *Ornithoptera rothschildi*, Rothschild's Birdwing, offer eye-watering flashes of jungle green and burnished gold set in dark shadow, like shafts of sunlight pooling on to the forest floor. Rothschild had financed the expedition to the Arfak Mountains of New Guinea where the butterfly was first discovered. He loved big, colourful butterflies as he loved big, bizarre animals (he stocked his park at Tring with his favourite bird, the cassowary, at least until one of them attacked his father). He himself named the world's largest butterfly, *Ornithoptera alexandrae*, after Queen Alexandra. As for the eponymous moth, it is an entire genus of silk moths, *Rothschildia*, each one with dazzle-patterned wings of pink, crimson and cream with translucent 'windows' whose shapes suggest the obsidian daggers of the Aztecs. The foretips of these moth wings are spotted to resemble the heads of snakes or even (it has been suggested) baby alligators. The colours of *Rothschildia* moths make you think of rich fabrics, heavy silks and regal robes, and then of blood and flickering fire. In their opulence that tips over into the bizarre they are apt metaphors for an extraordinary man.

Walter Rothschild stood six feet three inches high in his socks and weighed twenty-two stone. His great-niece Hannah remembered him as a 'huge, strutting bear of a man'.[17] Miriam recalled how he would bowl across Tring's marble hall breathing heavily 'like a grand piano on castors'.[18] At night his elephantine snores reverberated down the corridors. He was in fact noisier by night than by day for Walter Rothschild normally said very little, and when he did speak he did so slowly with frequent long pauses, looking at the ground for inspiration. Conversation with him was difficult verging on impossible, particularly as he also suffered from a crippling lack of voice control so that his long silences were broken by roars. He experienced difficulty turning his thoughts into words. His family came to realise that Walter was hopelessly unsuited for business. In Miriam's words, 'his anti-talent for finance, his stubborn silences, his slowness and blank spots and phobic interest in natural history' qualified him for one thing only but in that one thing he was supreme. That thing was his private museum.[19] There his multiple dysfunctions could be trumped by his 'peculiar brand of enthusiastic megalomania'. Necessarily Walter had a prodigious memory as well as sufficient (but not limitless) wealth. By the end of his life his museum contained the largest collection of animals, birds and insects ever assembled by one man. An itemised account would run into almost astronomical figures. It would include 2.25 million set butterflies and moths (but 'no duplicates'), 300,000 bird skins, 200,000 bird's eggs and 30,000 scientific books. Between them Walter, his brother Charles, his curators and his assistants, named 5,000 new species of insect, many of them butterflies, and published over 1,200 books and papers based on the collections.

The Rothschilds, both in France and England, were eager and indulgent collectors, filling their houses with precious and rare objects. Baron Henri, who, like Walter, was an animal lover, rather ironically amassed a famous collection of *têtes-de-mort*. Baron Edmond amassed precious engravings and stones; Baron Ferdinand went in for *objets de vertu* (that is, upmarket *objets d'art*). But Walter eclipsed them all in the extent of his monomania. His collecting began in early childhood. Like his brother Charles his original passion was for beetles. Then, abruptly and without explanation, he shifted to butterflies, perhaps, as Miriam suggests, because they afforded greater opportunities to study evolution. Tring Park house began to fill with butterfly cabinets, glass-fronted exhibition cases and expensive volumes on butterflies from around the world. Walter was eventually given a family allowance that left him free to do more or less as he wished. He never married, and gave up most of his other responsibilities in order to settle happily into a fixed routine 'of a fourteen-hour day of virtually incomprehensible

work – technical descriptions of the tail-end of insects – punctuated at intervals by flaming rows'.[20] He bankrolled what was
probably the most intensive phase of butterfly hunting ever. At
the peak of his activities Walter had collectors everywhere; a
contemporary map marking the whereabouts of Rothschild's
collectors with red dots looks as though the world has caught
measles. Consignments of exotic butterflies, each carefully
preserved inside layers of tissue, made their way to Tring from
all corners of the earth, ready for the lenses and microscopes of
Walter and his ever-busy curator of insects, Karl Jordan. Unlike
ordinary mortals who queue to have their papers peer-reviewed
and wait their turn for publication, Rothschild had his own journal,
the *Novitates Zoologicae*, to record the formal details and descriptions of each discovery.

Walter's personal collecting was confined to Europe and
North Africa; he never did see any of his favourite birdwings
in their natural habitat. Rather they were collected for him by
hardy adventurers such as the American Dr William Doherty
(1857–1901) who risked awful weather and bouts of fever to
camp out in the forest, bargain with the natives, shoulder his
gun and set his traps. Men like Doherty were happiest when
far from civilisation even though it could – and in his case did
– shorten their active lives. Collecting in rainforests undermined
the best constitutions. Doherty became increasingly gloomy,
nervous and fatalistic, lost his *joie de vivre*, and finally died of
dysentery, aged 44.[21]

Another of Rothschild's collectors, Captain Herbert Cayley-
Webster, claimed later to have fought off 'shoals of canoes full
of myriads of cannibals', leaving some of his men dead on
shore. All night long, he was kept awake by the 'horrible noise

of the banqueting gongs', while glimpsing in the flames the dark shapes of the natives as they barbecued the corpses of his friends.[22] But perhaps the most successful was Alfred Stewart Meek (1871–1943). It was Meek who secured the first specimen of the giant Queen Alexandra's Birdwing, some nine inches across the wings, by blasting the soaring, high-flying butterfly with buckshot. The specimen was too full of holes to send to Lord Rothschild but the resourceful Meek managed to locate its eggs on trailing vines of *Aristolochia*. Knowing he was on to a winner, he managed to rear a series of perfect specimens that made him a handsome profit. Collecting butterflies in the then virgin forests of New Guinea was difficult and sometimes dangerous work. Some of the most desirable species needed to be attracted down from the treetops with bait. Meek learned which stinky messes attracted a particular butterfly and which ones could be mesmerised by coloured sheets of paper or even a dead fellow butterfly pinned to a bush. He paid native tribesmen to help find the best spots and to collect butterflies for him. One of them shot the first specimens of the resplendent birdwing *Ornithoptera chimaera* with poisoned arrows used to hunt birds of paradise.[23] In the end Meek saved enough from his butterfly business to buy a cattle station in Queensland and later to retire to a villa overlooking Bondi Beach.

What he could not get Meek and others to catch for him, Lord Rothschild bought. He was a compulsive shopper. He combed the auction rooms for insect treasures and paid small fortunes to high-end dealers such as Le Moult of Paris (the original of the butterfly-collecting character in Henri Charrière's book *Papillon*, the name of which is French for 'butterfly'). Rothschild was considered the best, steadiest, and, of course, richest, of customers. Often, on receiving some interesting material, the trader would simply send the whole lot to him and invite him to take his pick. Walter Rothschild's method might seem greedy, like a spoilt child heaping up goodies, but he collected with purpose. He took only what he wanted for scientific comparisons and to elucidate forms and races of butterflies. Today everyone knows the word 'biodiversity' but it is close to becoming an abstraction; no more than a vague Noah's Ark sense of the variety of life. To get an idea of what it really means there is no better way than to examine a large collection of butterflies. There is evolution on their wings, history in their colours, all laid out for convenience in a cabinet drawer. If you can get over an understandable antipathy for dead insects and allow the magic to work it may open a window in your mind. As Miriam Rothschild (who knew Walter's collection well) put it, 'suddenly the outlook broadens, the horizon expands – a penny drops, new ideas materialise, the mind "takes off".'[25]

What saved Lord Rothschild's vast undertaking from the common fate of collections – broken up, auctioned and forgotten – was his decision to leave it to the nation, or as much of it as the nation wanted. Parts, including the British collections, are now housed in the Natural History Museum's new Darwin Centre. 'Its importance,' writes Miriam Rothschild, 'lies in the unfolding and presentation before your eyes of a whole order – in all its variety and complexity, culled from continent to continent, from one far flung oceanic island to another, from desert and forest and prairie and mountain range. There is also an indefinable factor about these collections, a Walterian factor – call it what you will – a whiff of zest and wonder, which must somehow have been pinned in among the butterflies.' Call it what you will; I think I will call it the Rothschild effect.

7.

The Golden Hog or The Wonderful Names of Butterflies

Have you ever wondered how the word 'butterfly' came about? What, after all, do butterflies have to do with butter? There have been various tentative explanations but few of them are very convincing. One supposes that the original 'butter-fly' was yellow, as yellow as a pat of fresh butter, and so was presumably the bright yellow Brimstone – strictly speaking the *male* Brimstone since the paler female looks more like margarine. But why should the Brimstone be the progenitor of all butter-flies? Yellow is not the principal butterfly colour; far more of them are shades of brown, blue or white. Another theory suggests that 'butter' comes from the Saxon word *beatan* meaning 'to beat'; it is the insect that flies by beating (rather than buzzing) its wings. You sometimes hear the suggestion that 'butterfly' got mixed up with 'flutterby'. But 'flutterby' is a modern word whilst 'butterfly' is unfathomably old; word researchers have proved that 'butterfly' precedes 'flutterby' by at least a thousand years. Yet another proposal is that the orig-inal name must have been 'beauty-fly'. Again, this kite doesn't fly. The 'butter' in 'butterfly' does not mean beauty. The one thing we can be reasonably confident about is that 'butter' means butter and 'fly' means fly.

According to the *Oxford English Dictionary*, the word 'butterfly' originated as the Old English *buttorfleoge* in an Anglo-Saxon manuscript written about 1,300 years ago.[1] It was probably a familiar word even then for the name is common to several North European languages. It is *botervlieg* in Old Dutch and *buttervleige* in Old German. A less complimentary name in Dutch is *boterschijte* – or 'butter-shit'. This is a cruel libel because butterflies hardly 'shit' at all; most of their diet of glucose leavened with salts is converted into energy; it is their caterpillars which are the notable, indeed stupendous shitters. But unless they are very sick, caterpillars' droppings are hard and friable; they do not resemble butter.

Beetles bite and spiders spin; their names mean, essentially, 'biter' and 'spinner'. Moths are named, rather unkindly, after their 'maggots': the little maggots that supposedly munch our woollen socks and pullovers; moths are 'munchers'. But with butterfly we need to forget about biting and munching and think instead about how ancient tribes might have regarded these bright-winged insects. A clue may lie in another old German word for butterfly, *Schmetterling*. It comes from *Schmetter*, a dialect word for cream, and hence the word takes on a similar meaning to butterfly: the 'little cream-fly'. Even more suggestive is a second folk name from central Europe: *Milchdieb* or 'milk-thief'. Stealing milk or whey is one of the things which, according to Teutonic myth, witches did; they robbed the worthy farmer by tweaking the udders of his cows

during the night. Is it possible that butterflies were once associated with milking, or perhaps attracted by the scent from the
butter churn? There are accounts from traditional farms in
Eastern Europe of white butterflies fluttering over the milk
pails. Possibly they are attracted by some pheromone in milk
– or perhaps by its colour, just as Purple Emperors are attracted
to reflections in puddles or on car roofs. We know that the folk
image of butterflies was not always positive. They were, for
example, seen in some country districts as bringers of bad luck.
One folktale holds that a white moth escapes from the mouth
of a witch to carry on her business while she sleeps.[2] Such
stories are faint echoes of a time when the natural world must
have been full of superstition – and fear.

We have the advantage in knowing that, apart from a few
crop pests, butterflies are harmless to mankind. The world
over, butterflies have impressed people by their airy, graceful
flight and brilliant wings. But they were, by the same token,
the closest things in nature to how people imagined fairies or
spirits. Dragonflies, which most of us now find equally beautiful and attractive, were regarded well into the twentieth
century as agents of the devil.[3] They looked like stings on
wings, and their huge eyes seemed capable of peering right
into your soul. Similarly moths, which include some of the
loveliest insects on earth, have been tarred and feathered by
biblical allusions to their fretting and gnawing; they were seen
as the insect equivalent of rust. Spiders, too, have a wholly
undeserved evil reputation that clings to them even today.
Perhaps we should see the word butterfly in the same light.
They were not 'flutterbyes' or 'beauty-flies' or butter-yellow
heralds of spring. They were, rather, sinister spirit-like beings

which showed an unhealthy interest in milk from the dairy yard.

The English names of butterflies and moths are unusually poetic. But they are sometimes also rather obscure. Who now would invent words such as 'admirals' or 'arguses' or 'hairstreaks', or describe a particular butterfly as 'clouded' or 'silverwashed'? We take these names for granted and do not consider their oddness. Butterfly names are often described as Victorian. In fact, they are older than that. What is easy to overlook is how apt they are.

Take 'hairstreak'. Nowadays we would probably say 'hairline' but that misses the vibrancy of the word 'streak', suggesting, to my ear at least, a flicker, like lightning, or the irregular motion of a pale line across a dark background. The irregular white line on the hindwings of these small, mainly woodland butterflies does seem to twitch as the insect flexes its wings, an effect accentuated by the little tail on the end of each wing. One species, the White-letter Hairstreak, even has a wiggle to the streak of white as though someone has tried to paint a shaky 'w' drawn with the thinnest of brushes. It is tempting to suppose that whoever first came up with that word 'hairstreak' had watched the living butterfly in its natural habitat and seen for himself that characteristic shake of its wings.

When I first fell in love with butterflies and their funny names, I thought I understood at least one of them. That was 'fritillary'. My father, who was a keen gardener, told me what a fritillary was. It was a kind of lily, he said, with a hanging, bell-shaped flower. It was pinkish, with chequered markings. I found a picture of one, perhaps on a packet of bulbs, and noticed how the box-shaped flower shared the same pattern as the butterfly. But was the flower named after the butterfly or was it the other way round?

The original fritillary, or *fritillus*, was neither, it seems, a butterfly nor a lily, but a wooden or ivory box with a chequerboard pattern used for shaking dice.[4] It made a good rattle before sending the dice clattering across the gaming table. It was the pattern that was remembered and anglicised as 'fritillary', as another word for 'chequered'. It seems a happy choice of group name for the bright, elegant butterflies with black markings on an orange-brown background, capturing something of their grace and airy dalliance. When I hunted them as a boy we called them 'frits'. It rhymes with 'flits' and that seemed all right too.

I also managed to work out for myself who or what 'argus' was. At that time I had never seen a real live Brown Argus butterfly, nor the unrelated Scotch Argus, but I knew that the model for both was a slightly scary figure from the Greek myths, a kind of super-shepherd whose distinguishing feature was the numerous eyes scattered all over the top of his head, like grapes. These gave him all-round vision which proved a tremendous advantage in his shepherding. It also meant that he could work a shift system with some eyes closed and others wide open and alert. Butterflies, I knew, have only two eyes, like us, not the unique and, in evolutionary terms, highly improbable arrangement of Argus the Shepherd. But, in addition to its genuine pair, the little Brown Argus has numerous fake eyes sprinkled all over the undersides of its wings: little black dots each surrounded by a circle of white, six on the forward wing and twelve on the hindwing. These are its 'argus' eyes. Being no more than arrangements of scales they are of course blind but, you think, all those little staring eyes might well cause a moment of confusion to a predator, such as a hungry skylark. Alternatively, the confused bird might strike instinctively at one of the fake eyes instead of the real ones. A butterfly can get by with a tear in the wing, but a peck on the head brings down the curtain. Argus, I thought, is really a rather good name for this kind of butterfly. And a memorable one too: much better than 'Little Brown' or 'Dainty Dipper', which is probably what we would call it today.

By contrast, modern English names, of the kind coined for other insects such as ladybirds or dragonflies, are more pragmatic than poetic. They have little truck with art or mythology, and are invented not so much in a spirit of admiration as an aid to identification. In the case of dragonflies, many of their names are based on flight behaviour. One group are the 'skimmers', named from their habit of flying close to the water's surface; others are 'hawkers' which patrol the pond margin. The 'darters' skitter about over the water whilst the 'chasers' pursue their prey across the pond. If butterflies had gone down that route we might have gliders, soarers and flutterers.

Bumblebee names are even more boring. Many of them are named after the tuft of coloured hair on their rear ends – the 'bum' in bumblebee, so to speak – such as the Buff-tailed Bumblebee and the Red-tailed Bumblebee. The names of ladybirds are, if possible, even less imaginative, being based mainly on a spot count so that you start with the two-spot ladybird and end at the twenty-four-spot via most numbers in between. By the same token butterflies might have been named from their spots or tails.

It is often pointed out that butterfly names are unscientific. The Marbled White is not related to the Large White, the Red Admiral isn't a close relative of the White Admiral and the Brown Argus has nothing in common with a Scotch Argus. But what they have instead, and which is much more unusual, is 'cultural resonance'. The names of butterflies draw us into a world where art meets science, often producing names which scintillate in the mind and engage our feelings, a place denied to multi-spotted ladybirds or coloured bumble-bums.

It took a surprisingly long time for butterfly names to 'evolve'. Let us begin at the usual beginning, with the dazzled descriptions of the Tudor physician, Thomas Moffet. In his treatise on the 'lesser living Creatures', *The Theatre of Insects*, written around 1589, Moffet made clear his admiration for butterflies by devoting a whole chapter to them.[5] But although he managed to illustrate around two dozen kinds with crude woodcuts, he had no names for them. He had difficulty describing them, too, since there were no existing words for insect anatomy either. Instead Moffet drew on his knowledge of birds and beasts so that his butterflies have a 'belly' as well as a 'snout' and a 'beak', and also 'horns' or 'cornicles' since their antennae stick out like the horns of a bull.

Moffet knew, or thought he knew, a few hard butterfly facts. He knew that some kinds lived longer than others and that they approached winter in a 'languishing condition'. Tortoiseshell butterflies survived the coldest months for Moffet had found them in his house, 'sleeping all the winter like Serpents or Bears, in windows, in chinks and corners'. He believed the old story that certain large moths attack sleeping butterflies by night, beating them with their wings 'as great Tyrants devour and spoil their subjects'.

Without names to give them, Moffet drew on his literary gifts to try to do justice to their colours and patterns. In the wings of his Peacock butterfly, for example, he saw 'four Adamants [i.e. diamonds] glistering in a bezel of Hyacinth' which 'shine curiously like stars, and do cast about them sparks of the Rain-bow'. You could regard it as 'the Queen or chiefest' of all the butterflies, he suggested.

He was less enamoured of the Painted Lady, which lacks the Peacock's iridiscence, though he did note its resemblance to the skin of ladies who spend most of their time indoors: 'Nature bred this [butterfly] with a chamblet-mingled coat [i.e. a garment made of fine cloth], but it wants lively colours, for the wings are of a black reddish fading yellow and russet colours, and it is more beautiful for its soft skin, than for its gallant apparel.'

He admired fritillaries, too, especially those with silver spots inset into the hindwing like pearls. One in particular 'holds forth a rare list of oriental Pearls shining in blue, the upper wings being of a flaming yellow, show[ing] like fire'. It is hard to be certain which one he meant but it might have been the Pearl-bordered Fritillary, once a common springtime butterfly.

It was in words like these, owing more to art than science, that butterflies first entered the consciousness of the curious-minded. Colourful, gem-studded butterflies such as Peacock, Small Tortoiseshell and the Pearl-bordered Fritillary showed off 'the elegancy of Nature'. But since almost nothing was known about them, the best anyone could do was, in Moffet's words, 'to admire the work of a bountiful God, the author and giver of such rich treasure'.[6]

Awkward labels, rather than names, also mark the work of the great naturalist John Ray (1627–1705), the Essex parson-naturalist who set about ordering the various kinds of insects, although he never got much further than the butterflies and moths. He realised that without names it was impossible to talk intelligently about butterflies. He knew only a few: country names like fritillary, tortoiseshell, 'painted lady', 'peacock's-eye', and, more surprisingly, 'small heath'. But Ray preferred the device of short descriptive tags in Latin which were as learned as they were impracticable. Nor were they particularly imaginative. His tag for the Marbled White, for example, trans-lates as a 'Butterfly of middle size with wings beautifully vari-egated with black and white'.[7]

No one was going to call it that. Fortunately there was at least one contemporary who saw the need to invent simple names where none existed. We have met him before. He was the collector of squashed butterflies, the patron and friend of Elinor Glanville, James Petiver.

Petiver needed names. He made a practice of publishing short descriptions and simple engravings of butterflies and moths in self-published catalogues he called Gazophylacia (or 'chests of valuables'). It seems to have been Petiver who gave us the basic vocabulary of butterfly names: 'hairstreak', 'argus', 'fritillary', 'brimstone', 'admiral' and 'brown'. Not many of his full names have stood the test of time, and anyhow he kept changing them, but Petiver deserves the credit for at least the idea of a proper name, in English, for every species of butterfly.[8]

Some of his early efforts were close in spirit to Ray's descrip-tive tags, such as his 'brown-eyed-Butterfly with yellow Circles' (the Ringlet) or 'small golden black-spotted Meadow Butterfly' (the Small Copper). Petiver also had a habit of naming a species after the person who brought it to him, such as 'Handley's brown Butterfly', now better known as the Dingy Skipper. Others he named after the place where they were first found, such as Enfield Eye for the Speckled Wood or Tunbridge Grayling, soon to be simplified to Grayling. His first specimen of the beautiful Swallowtail was 'caught by my ingenious friend Mr Tilleman Bobart' and named 'The Royal William' after the reigning king, William III, perhaps because it had been caught in the garden at St James's Palace. Only half a century later, when memories of King Billy were dimming, was the name changed to the Swallowtail.

Of all Petiver's forgotten names, the one I most regret is 'hog' (he seems to have pronounced it 'og') for the small, chubby butterflies now known as skippers. As the most moth-like of butterflies, with their broad 'faces' and little black eyes, they do conjure up images of fairy pigs, especially when the fawn blur of their wings preserves the plump body in sharp focus, like a hummingbird. Petiver knew only two kinds of hog, his large 'Cloudy Hog' and the slightly smaller and brighter 'Golden Hog'. Later on they were renamed the Large and Small Skipper respectively, after their characteristic skipping flight. Perhaps 'skippers' seemed more dignified than 'hogs'. And so, alas, we lost our little flying pigs.

By the time *The English Moths and Butterflies* by Benjamin Wilkes was published in 1748, many butterflies had acquired the names by which they are known today, including Clouded Yellow, Orange-tip, High Brown Fritillary and Purple Emperor. All these names incorporate a colour, and it might be significant that Benjamin Wilkes and many of his fellow 'Aurelians' were professional artists; his card describes him as a painter 'of History Pieces and Portraits in Oyl'.[9] It was surely an artist's colour sense and imagination that produced such timeless names as Clouded Yellow and High Brown Fritillary ('high' meant richly coloured, and not a reference to high-flying habits). The name Silver-washed Fritillary so perfectly describes the pearlescent smears and floods of silver on the butterfly's hindwings that you can almost see Wilkes's brush reproducing the effect.

Moses Harris (1730–c.1788), who painted perhaps the most beautiful of all historic butterfly pictures, seems to have had a particular talent for names.[10] It comes out most clearly in those he coined for moths which include such novelties as the Merveille du Jour and the True Lover's Knot. Harris is exceptional in that he sometimes explains what he had in mind. The butterfly hitherto known as the London Eye or Great Argus, for example, he renamed 'the Wall' – not so much because of its wing pattern, even though it *is* rather brick-like, but because 'it frequently settles against a field bank, or perhaps against the Side of a Wall; and is for this Reason, called THE WALL FLIE'. He also introduced the Gatekeeper as an appropriate and pleasing name for a butterfly that loves to fly along 'the sides of hedges in lanes and meadows'. Others with the same ring of originality include the Speckled Wood, the Silver-studded Blue ('stud' is exactly right for its tiny, diamond-bright twinkles) and Camberwell Beauty. Harris also gave us the Duke of Burgundy but in its case he unfortunately forgot to tell us what he meant by it.

Hence most of our butterfly names are not Victorian. They are from the century before Victoria and so broadly Georgian – or in a few cases older still: Williamite or 'Annian'. By the end of the eighteenth century the poetic flights of earlier times were receding. The last of the classic Georgian butterfly books, William Lewin's *Papilios of Great Britain*, gave us the Large Blue and Small Blue – dull names compared with 'Adonis' and 'Silver-studded' blues. Lewin also changed one of Harris's compositions, the 'Dishclout or Greasey Fritillary' to the more prosaic Marsh Fritillary. The chequered butterfly hitherto known as 'Vernon's Half-mourner' – from a form of mourning dress that allowed white as well as black – became the Bath White. Lewin's explanation was that a young lady from Bath had commemorated that butterfly in a piece of needlework. But anyone visiting that city hoping to see a Bath White butterfly was in for a disappointment.

The probable reason why we still use butterfly names that are upwards of 200 years old is through two pieces of good fortune. The first was the publication of *Lepidoptera Britannica* by Adrian Haworth in the opening years of the nineteenth century. This became the standard text on the classification of butterflies for the rest of the century and, as it were, fixed the English names in aspic (though Haworth would much rather you used the Latin names of Linnaeus). The few new butterfly discoveries of that century slipped naturally into the established system: Scotch Argus, Mountain Ringlet, Black Hairstreak, Essex Skipper. Haworth himself made a few judicious changes: he preferred Chequered Skipper to Spotted Skipper and, less happily perhaps, Silver-spotted Skipper for what had up until then been the Pearl Skipper.

The second piece of good fortune was the long reign of the popular *Butterflies of the British Isles* (1906) by Richard South which was in effect the standard work for seventy years.[11] Its author had the conservative disposition and good sense to stick with the names he had inherited. And no one has seriously sought to change them since then. It seems we like our strange, allusive butterfly names. We may not think about them much – entomologists are not, in general, connoisseurs of language – but unlike modern names they bear witness to how people felt about butterflies. The spirit of the old forgotten pioneers lives on in them, even when we regard them as nothing but labels.

NYMPHS AND SHEPHERDS COME AWAY

For a long time you could expect to lose credibility in the eyes of fellow entomologists if you dared to call a butterfly a Clouded Yellow instead of *edusa*. Latin names were the only scientifically correct names; the English ones were strictly for non-scientists. Yet even to scientists, Latin names do not mean much. Few, if any, butterfly books devote space or effort into explaining what they mean and why the species were so named. But they at least meant something to the person who invented them – and unlike the English names, we know who those people were for their surnames, in abbreviated form, follow the binomial scientific name. And when you delve into them you find, more often than not, that it is not impassive science but the same romantic sensibilities at work, deftly disguised within a learned language. You discover analogies with nymphs and shepherds, satyrs and devils, mirrors and jewels, even thoughts and dreams. Latin, quite as much as English, is part of the cultural identity of a butterfly.

By custom, the larger, more colourful butterflies, especially those in the Nymphalidae family, are named after female characters. Most of them are not historical but princesses or minor female deities taken from Ovid, Virgil and other classical poets. Like butterflies, nymphs were supposed to be beautiful and graceful and bring joy into the lives of others. And they were associated with springs, flowery groves and mountain pastures – which are, of course, good habitats for butterflies. For example, the Peacock, *Inachis io*, is named after Io, a beautiful girl who was seduced by Zeus and then turned into a heifer to prevent his wife taking revenge (Inachis was Io's father). The White Admiral, *Ladoga camilla*, commemorates Camilla, a warrior princess from Virgil's *Aeneid* (did the name 'admiral' require a militant lady?). *Cynthia*, the name of the Painted Lady, is not so much a character as an inspiration, a muse. The name was a popular one among eighteenth-century lyric poets. It was also an alternative name for Diana the huntress, the goddess of the great outdoors.[12]

The most flattering names were reserved for the fritillaries. They include the Three Graces. The Pearl-bordered Fritillary is *Euphrosyne*, bringer of mirth and joy, perhaps a coded reference to how one feels on spotting this bright butterfly on one of the first warm days of the year. The second is *Aglaia*, the embodiment of beauty and splendour, a name appropriately awarded to the Dark Green Fritillary. The third of the Graces, Thalia, had already been used for another insect. But all that was needed to stay within the rules was a slight modification and so it was as *Athalia*, the Grace who brings music and song, that the Heath Fritillary was named. Perhaps bursting into song was the natural response to spying the rare Heath Fritillary.

The browns, as suited their more sober colouring, were seen as masculine personalities. They were the Satyrs, goat-footed beings that lived in woods. In keeping with their dark colours, these butterflies sometimes embody shady or unlucky characters. Such is the Gatekeeper's name of *tithonus*, named after a foolish youth who begged the gods for everlasting life only to discover that it was a curse. What he got was not eternal youth but eternal decrepitude; wizened and perhaps browned with age he prayed now for death as he was slowly consumed by, in Tennyson's words, 'cruel immortality'.

Another gloomy name is *Maniola*, the genus containing the Meadow Brown and meaning 'the little shade of the departed' or, as we might say, 'the mini-ghost'. The butterfly's dusky wings suggested the murk of the nether regions where departed souls dwelled. The same thoughts were aroused by even darker butterflies that fly in Europe's mountains, species of the large genus *Erebia*, named after Erebus, the mythical region of darkness beneath the world. One of the two British species, the Scotch Argus, *Erebia aethiops*, takes its other name from the Ethiopian, then a general name for the dark-skinned race of humans. The other, the Mountain Ringlet, *Erebia epiphron*, is based on a Greek word meaning 'thoughtful'. That too was suggested by the melancholy hues of the butterfly, though separating it from some very similar-looking European species does indeed require thought. Nor is it a cheerful thought that this, our only true mountain butterfly, may soon become a victim of climate change.

An exception among the browns is the Grayling which is named after a female character, Semele, a mortal beloved of the gods and best known from Handel's opera of that name. Appropriately to our theme, Semele's story is a sad one for she perished in the ensuing smoke and flames once Zeus revealed himself to her in all his glory. Perhaps it is only coincidence but the male Grayling does have a smoky look, while its fondness for settling on paths puts you in mind of the best-known aria from that opera, 'Where'er you walk'.

Anyone who has watched a skipper will know that its flight is not fluttery like other butterflies but a blur of wings, almost a buzz, and more like a moth than a butterfly. The movements of the Small Skipper reminded the entomologist Jacob Hübner of dancers in ancient drama who would hop and skip about the stage, pausing now and again with their hands spread wide in a gesture of amazement or sorrow. Such actors were called the Thymelicos, and so, following that, the skipper's name *Thymelicus sylvestris* means 'the little dancer of the woods' (there was obviously some muddle here since the Small Skipper actually prefers rough, unshaded grassland). The Large Skipper, on the other hand, flies up from its perch to evict an intruder, returning to its place of rest much as a flycatcher does. Again, Hübner had a name for it: the butterfly became *Ochlodes*, meaning turbulent or unruly, a reference not only to its agitated flight but also to its 'character'. Jeremy Thomas described the Large Skipper as a 'burly little butterfly, darting in golden flashes'.[13]

A more sinister name, *Erynnis*, was reserved for the Dingy Skipper which has the habit of flying up from the path with a movement compared by Thomas with 'aircraft peeling off from a formation'.[14] To another German entomologist, Franz Schrank, this restless behaviour suggested the Furies, the 'Erinyes', who pursued wrongdoers, hounding them from place to place until their victims were driven mad. In Schrank's poetic imagination, that was the fate of the Dingy Skipper, for ever pursued by invisible avengers. Its species name, *tages*, commemorates the same behavioural quirk. In myth, Tages was the boy with the wisdom of an old man who *suddenly rose from the ground* to instruct the Etruscans in the art of divination.

The Wall is another butterfly that likes to sun itself on the bare soil of footpaths, and it too has a blood-curdling name. It is *megera*, after Megaera, one of the Furies, who, full of envy and spite, fastens her attentions on adulterers. Its modern genus name is more benign. It is *Lasiommata* or 'hairy eyes', a reminder that butterflies were described from dead, pinned specimens in a collection, for, although the Wall does indeed have 'eyelashes' they are difficult to see without a hand lens.

The Danish entomologist Johann Christian Fabricius (1745–1808) was the first person to arrange butterflies into families of related species. He was also the first to distinguish clearly between butterflies and moths. Many of the Latin names ascribed to our butterflies originated in his terse descriptions (themselves often based on collections in London). 'Fab' was, it seems, fond of puns and analogies, often obscure ones. He might have missed his vocation as a compiler of cryptic crosswords. Take his teasing name for the Purple Emperor, *Apatura iris*. Iris presents no difficulty: she is the personification of the rainbow, an obvious allusion to the male butterfly's dazzling purple iridescence. But *Apatura* is a real puzzle; evidently a made-up word, possibly an anagram. The most likely explanation for it is that Fabricius had seized on the Greek word *apatao*, meaning to deceive. The Emperor's mantle of purple is a question of 'now you see it, now you don't'; one moment the butterfly is sombre brown and the next, an eye-wateringly brilliant purple. But

Fabricius, too, has become a deceiver by turning the original word into an invented one of his own. He was, you could say, identifying with the butterfly, making a learned little joke which hinges on the idea of deception. For all we know, it might have seemed brilliantly funny in eighteenth-century Denmark.

The butterflies one feels most sorry for are the smallest ones – the blues and skippers – which the great Linnaeus relegated into his sixth and last category of butterflies, *Plebejus*, the plebs. They were the commoners of the butterfly world, the little workers and peasants, not much bigger than clothes moths. Later on the insult deepened when the tribe of blues was renamed the *Plebejus parvi*, the *poor* plebs; even the two cabbage whites were deemed to be in a higher class. This system of classification has long since been consigned to the dustbin but the plebeian label lives on in *Plebejus argus*, the scientific name of the Silver-studded Blue: a butterfly whose beautiful silver 'studs' seem to bely its implied lowly status.

As some compensation, several of our smallest butterflies have particularly beautiful names. It seems fitting, somehow, that the smallest of them all, the Small Blue, *Cupido minimus*, is named after Cupid, the fluttering love baby with his bow and arrow. And among the most unexpected is the Green Hairstreak's name: *Callophrys rubi* or 'beautiful eyebrows in the bramble'. Close up you can see why: above each dark eye there is a scattering of iridescent scales, like the psychedelic glasses worn by Elton John, while the eye is further outlined in white as though the butterfly were wearing make-up.

8.

Seeing Red

The Admiral

Of all the colours of the rainbow the one that makes the greatest impact on the eye is red. There is something instantly arresting about the colour of fire and danger. It is charged with alarm and urgency. We have red fire engines, red warning lights and the Red Cross. If we want something to be seen from a distance, such as a post box or (formerly) a telephone kiosk, we paint it red. Perhaps it is because red is so conspicuous that few wild animals are truly red: even the so-called Red Squirrel and the Red Deer are really shades of reddish-brown that conceal rather than advertise.

Relatively few butterflies are red either. In Britain, and indeed, in all mainland Europe, only one species has a pattern of full-on bright red, like a splash of blood: the Red Admiral. Its redness is, if anything, emphasised by its contrasting colours of black (or nearly black), white and a powdering of reflective blue. In flight that bright red power-band seems to flash on and off like the lights of an ambulance. The bold colours of what Nabokov called this 'magnificent, velvet-and-flame creature' make the Red Admiral instantly recognisable. Everyone knows it and, it seems, everyone always did for it is by far the most frequently reproduced butterfly in art. You find it perching and fluttering in Dutch flower paintings of three or four centuries ago, and before that in the margins of medieval psalters and books of hours. The Red Admiral has always been a butterfly that attracts notice.

It is also a very successful butterfly. As a long-distance migrant it is one of the world's most widespread species. In Britain, as in continental Europe, it is a common and regular visitor to gardens, especially in late summer. While the Red Admiral's attention is focused on sucking the juices from an ivy blossom or a rotten apple it ignores us and we can approach it closely. It is then that you can appreciate the power hidden within that slim, apparently fragile body. The thorax, the bit that powers the wings, is leathery, a pouch strapped with tough ligaments that beat the wings on the Admiral's tireless journeys over land and sea. Watching its L-shaped proboscis quivering inside the nectary of a flower, it reminds you not so much of a dancing butterfly as the fuel intake of an engine, a light aircraft perhaps, with high-octane fuel pouring into that miniature carburettor.

Since redness is this butterfly's most obvious attribute, why was it named after a senior naval commander? The fact that this butterfly Admiral can and does cross the sea is irrelevant; it was so named long before its tremendous migratory journeys were known about or even suspected. The usual explanation is that 'Admiral' is a corrupted form of 'Admirable'. That, they say, was the original name: the Red Admirable, the admirable Admiral.[1] Neat and altogether apt, it is an explanation that satisfies and looks like a fact. Indeed it is repeated over and over and has acquired all the properties of a fact except one: it isn't true.

The 'Admiral' is one of our oldest butterfly names. It appears in texts dating back to the early 1700s. Quite possibly it is a folk name and, if so, it is one shared by other North European countries, notably Germany and the Netherlands (which, it may be recalled, also share the word 'butterfly'). Fortunately we know roughly what 'Admiral' means from a chance remark by the eighteenth-century naturalist James Petiver in one of his illustrated 'treasure' catalogues. The analogy is not, it seems, with the naval commander himself but with his flag.[2] Petiver used the word Admiral for butterflies with broad wings on which the colours are confined to the corner leaving the rest plain – rather like the British Red Ensign. He called several non-European butterflies by that name, but the godfather of Admirals, his original Admiral, so to speak, was the Red Admiral, the brightest and boldest of them all. Petiver was writing at around the time of the Act of Union of Scotland and England in 1707 which resulted in a new national flag: the Union Jack. Whether or not by coincidence the Union Jack – and the Dutch flag too – flies the same colours as our butterfly: red, white and blue.

By the time Linnaeus was creating his system of binomial Latin names for animals, insects and plants in the 1750s, Admiral was this butterfly's common name throughout northern Europe. Linnaeus, too, knew it as 'Admiral', although, as a good international scholar, he latinised it to '*Ammiralis*'. Only the French had a different name for it: *Le Vulcain*, from Vulcan, the blacksmith of the gods, working his dark forge with its flickers of red fire and blue-white hot iron. In parts of England it was also known as the Alderman, possibly because civic robes were of the same colour, or perhaps because Alderman and Admiral share similar letters of the alphabet. But many authors preferred to call it the Admirable, believing, or wanting to believe, that this was the true name.

Less well-known is the Red Admiral's scientific name but it too is a memorable one. It is *Vanessa atalanta* (it was the one 'Latin' name I knew as a small child, though I probably said, 'atlanta', not 'atalanta').

Vanessa atalanta was a name worthy of the butterfly and it came from the heart. It was another of the cryptic names of Johann Christian Fabricius. 'Fab', as he appears in abbreviated form after his scientific names, was an anglophile. He visited London repeatedly to view its famous insect collections and made many friends there. Perhaps it was in token of the hospitality he received that he gave this favourite among butterflies an 'English' Latin name: *Vanessa*. Fab knew and obviously admired the work of the Anglo-Irish writer and satirist Jonathan Swift, for the original Vanessa comes from his epic poem *Cadenus and Vanessa*. Written in 1713 but not published for another thirteen years, it is an autobiographical love poem dressed up as a fairy story of nymphs and shepherds. Behind it lies the real-life Lolita-like story of the affair of the middle-aged Swift and his teenage pupil, Esther Vanhomrigh. Vanessa was his pet name for her, made by combining 'Van' with 'Esther'. Also in his mind, perhaps, was a pun on *Phanes* , of which *Phanessa* is the female derivative, the Greek deity of procreation and the generation of life. The word *phainos* also means 'one that shines'; it is one of the French names for a moth, *phalène* – a reference to how moths seem to reflect light from a window or torch beam. Poor Esther/Vanessa died young, aged 35, and it was only then that Swift published the poem. In his mind, and perhaps also that of Fabricius, the name Vanessa combined notions of beauty, brilliance and also the poignancy of a young life struck down in the midst of its flowering beauty, like a butterfly. You can imagine Fab reading the poem and pouncing on one particular line: 'When lo! Vanessa in her bloom/Advanced, like Atalanta's star'. There, ready-made, was his name: *Vanessa atalanta*.[3]

So who was Atalanta? In the eighteenth century, when every young scholar spent hours and hours translating Virgil and Ovid, she was a well-known mythical figure. The star of one of Handel's less-known operas, she was a legendary huntress and athlete, holding the mythical equivalent of Olympic gold medals for running, archery and throwing the javelin. She was beautiful but she was also, as a priestess to Artemis, untouchable: a dangerous woman to know. Atalanta's myth resounds with blood and fire. Fed up with her pestering suitors, she challenged them all to a race to the death. The plucky runners set off, were all overtaken by the fleet-footed Atalanta, and shortly afterwards their severed heads were dripping blood from poles set up around the arena. The only man to win a place in her chilly affections was a fellow sportsman, Meleager, but he too came to a sticky end, perishing in flames after offending a goddess. The elements of love and longing, blood and fire, all commingled in *Vanessa atalanta* must surely have given Fabricius a private satisfaction. Nor is that kind of sensibility necessarily dead. Vanessa is still a fairly common name, and I never think of the left-wing actress Vanessa Redgrave – 'Red Vanessa' – without wondering whether her father, Sir Michael, had been interested in butterflies.

THE BUTTERFLY FROM HELL

In invoking the fiery spirit of Atalanta, Fabricius might also have reflected on the folklore of the Red Admiral, which is full of foreboding. Past generations vested butterflies and moths with magic and myth. Even as the new rationalisations of science were changing our perceptions of life, there was a lively contra-flow of fear. The natural world might be full of useful plants and charming animals but it was also dangerous. The wings of the Red Admiral, so suggestive of flames in the dark, could act on an artist's imagination. A very early Red Admiral flies on the margins of the *Luttrell Psalter*, written and illuminated around 1330. It is being chased by a fantastical bird and the context is one of the Psalms of David, a prayer for deliverance from his enemies.[4] Presumably the butterfly represents the wicked, chased out by an angel in the form of a bird. Did the artist choose the Red Admiral because it already had that kind of reputation? Was it a suitable image for 'the bloody and deceitful man' of the Psalm?

A century on, a much more realistic Red Admiral flies above carnations and columbines in a portrait of a young woman by Pisanello made around 1440. The woman was probably Ginevra d'Este (the artist includes a *ginievre*, a juniper bush, in the portrait as an allusion to her name). The picture was painted after the death of the sitter, possibly from a death mask. Ginevra had died at the age of 21, reputedly poisoned, some said by her husband.

In the 'language of flowers', the garlands of columbines allude to death, the carnations perhaps to purity and innocence.[5] A butterfly, usually a white one, sometimes found a place in such a painting, perhaps as a sign of hope or promise of resurrection. But the choice of the Red Admiral suggests something else. Could it stand for the flames of hell, the fate that awaited her murderer?

The early Dutch master of flower painting, Ambrosius Bosschaert the Elder (1573–1621), often included a Red Admiral in his pictures. It almost becomes his artistic signature. His Admirals are usually found perched with partly closed wings on a stalk of a detached flower at some distance from the vase or floral basket. The wings are painted with the same meticulous detail that characterises all his work but the butterfly's body never looks quite right. The head in particular resembles that of a bird with the mouthparts distorted so as to resemble a beak. If his flowers stood for the beauty of life, did the butterfly symbolise death? We are left guessing, but the way Bosschaert's Red Admirals sit apart, in semi-darkness, perched on a withered blossom rejected from the arrangement, suggests as much. Everything in life has a counterpoint and the butterfly is perhaps the necessary corollary to the sweetness of the blooms. It is easy to suspect that its beauty is of the most sinister kind.[6]

There is no doubt about that in two slightly later paintings by Dutch masters. In *Madonna and Child with Two Butterflies* by the Jesuit artist Daniel Seghers, one butterfly is pure white and the other is the Red Admiral. The Virgin's gaze is directed to the white butterfly and the Christ Child's hand is raised as if to reach out to it. Yet, unexpectedly, his eyes are not on the white butterfly but on the red-banded one that flutters in the darkness beyond. The artist seems to have chosen two contrasting butterflies to make a moral point. The white one stands for the unblemished soul while the red is the embodiment of sin and temptation. The Admiral flutters on the far right of the picture which in triptychs of the Last Judgement is traditionally given over to hell. Seghers might be telling his Roman Catholic congregation that the Christian must strive for goodness but be aware of sin. Equally the white butterfly might stand for the unblemished Christ Child and the red one a foreknowledge of his death on the Cross.[7]

In the second picture, *Lobsters and Fruit* by Jan Davidszoon, painted around 1670, a Red Admiral flutters above a pile of mouth-watering food dominated by an enormous cooked lobster. The red of the butterfly matches the red of the lobster and that is no coincidence. The picture's theme is sin in the form of gluttony and the lesson is that greed leads to ungodliness. The great red lobster makes an effective stand-in for the Devil who, as everyone knew, wore red and bore a pair of lobster-like horns. The butterfly is a kind of Devil's second, an insect Mephistopheles, which helps to rub in the message. Lest there be any remaining doubt, there is an upset goblet, suggesting that the unseen glutton has had one too many and is now snoring under the table.[8]

Vladimir Nabokov recalled how vast numbers of Red Admirals had appeared over the steppes of Russia as far as the Arctic Circle in 1881, the year in which the Russian state was paralysed by the assassination of Tsar Alexander II.[9] The butterfly acquired a reputation as a messenger of death when people noticed wing markings which seemed to prefigure disaster. Low down on the hindwing is a distinct figure '8' and next to it a wobblier but still discernible figure '1' (it is clearer on some butterflies than others). Once you 'see' it, the spread wings spell out in full the fateful year 1881. The same arrangement of numbers inspired its Spanish name of *Numerada*. Henceforth the Red Admiral became known by superstitious Russians as the 'butterfly of death'. Peasants would cross themselves whenever Red Admirals appeared in large numbers. It was with that in mind that Nabokov has a 'dark Vanessa with a crimson band' settle on the sleeve of John Shade in his novel, *Pale Fire*. Any alert reader who understood the allusion would realise that John Shade was about to die.[10] Perhaps that is also why the film *All Quiet on the Western Front* ends with the soldier reaching out to a butterfly that had settled on the parapet of his trench, only to be shot through the head by a sniper.

The Red Admiral's sinister reputation does not seem to have followed it to England unless references to dark butterfly 'witches' in the north of England meant this species. We can enjoy the butterfly for what it is: a beautiful and welcome visitor which gorges on flowers and over-ripe fruit and lays its eggs on the humble nettle. But you are left wondering how its brilliant red bands evolved and what advantages they confer. The obvious deduction is that red is a warning colour. It is usually a sign that an insect is toxic and not to be eaten. The red Cinnabar Moth has such a strategy and it is indeed packed with poisons collected by its caterpillar from its famously toxic food plant, ragwort. But if that is the function of the Red Admiral's scarlet bands then it is a bluff. It feeds on non-toxic stinging nettles. As far as we know, the butterfly would make a tasty snack for a hungry bird.[11]

Sometimes an edible butterfly defends itself from predators by mimicking a toxic one. But, if so, what is the Red Admiral pretending to be? In Europe, at least, it does not resemble any other butterfly. How then, you wonder, did its elaborate patterns evolve: the scarlet band, those dapples of white and scribbles of powder blue and that mysterious triangle of yellow on the underside? A clue might lie in the pattern of its underwings, the ones that are visible when the butterfly is at rest and so at its most vulnerable. In the case of its relatives, the tortoiseshells and the Peacock, their undersides are dark and matt-coloured to conceal the butterfly. But the Red Admiral retains those bright bands on the undersides too; its camouflage is partial at best.

As Henry Walter Bates observed, the wing patterns of butterflies are nature's tablet: there is a story there waiting to be read.[12] In the case of the Red Admiral, an ingenious 'translation' was offered recently by Professor Philip Howse in his wonderfully oddball book, *Butterflies: Messages from Psyche*. Howse has gazed long and intently at the wing patterns of butterflies and moths trying to work out their biological message and meaning. In some cases a picture began to form in his mind, rather like those 3D arrangements of dots and dashes that suddenly reveal a donkey or a goldfish. What he suddenly saw in the Red Admiral nearly made him fall off his chair.[13]

The trick, according to Howse, is not to see the butterfly as we do but to imagine how it might be viewed by a predator such as a hungry bird. We know that birds 'see' in a different way from us; they do not need the whole picture but respond to certain patterns and colours. It is the key elements that matter, each one seen in isolation from the rest. As Howse puts it, birds, like autistic people, don't have 'Gestalt' perception: 'When a bird looks at a moth it sees the details first – the eye-spots, the waspish stripes or whatever happens to be there.'[14] It is such details, imprinted, we might guess, in the bird's memory, which determine whether a butterfly is treated as food or as something to avoid. In the case of the Red Admiral it could be the combination of colours that informs the bird rather than the red bands alone. What other animal shares these same basic colours, and in the same order? There is one: the goldfinch. There is the same slash of red around the bill, enclosing a dark beady eye that glints blueish in the light, then a band of white, followed by a collar and crown of black. Is it possible, then, that, for an instant, a feeding bird might mistake a resting but aroused Red Admiral for – a goldfinch?

Red Admirals often feed and roost in thickets of ivy, bramble and thistles, the sorts of places foraged by finches especially in late summer and early autumn. Imagine you are a foraging goldfinch poking about here and there with that sharp, triangular bill, when you spot something, a potentially edible morsel, roosting among the leaves. The butterfly raises its forewings with an audible rustle, and within an instant the finch sees a crude image, more or less life-sized, of itself, complete with that beady, blueish 'eye' and a blurry image of its sharp, triangular beak. The bird, you assume, would back off. The butterfly would live to fly another day.

Even that may not be the end of the Admiral's box of defensive tricks. Howse took another thoughtful gaze at the red bands on the hindwings, with their row of black dots and dark, scalloped edges, like little tube feet, and was reminded of a caterpillar with its row of dark breathing holes (spiracles), specifically certain toxic New World caterpillars that feed on poisonous vines. The same pattern is repeated, more faintly, on the underwings. The Red Admiral has a huge world range extending to the tropics of Central America. It has to find a defence to predators across a large slice of the planet. It may therefore pay to have more than one trick in its armoury.

Not everyone agrees with Howse and it has been suggested that the wing patterns of butterflies are there mainly to help the butterfly to identify its own kind. Personally, I believe that Professor Howse is on to something. There are, after all, moths that look like dangerous bees or hornets, caterpillars that resemble snakes, and even insects that, when looked at from a certain angle, suddenly seem to turn into toads, owls, or, in one case, a small but leery crocodile. If their evolutionary purpose is to create doubt and confusion in the mind of a predator, then the crude face of a goldfinch in the wings of a butterfly might perform the same trick. But whatever evolutionary route produced those glorious red wings, the result has been a resounding success. It is pleasant to think that those scarlet flashes that have engaged artistic imaginations for so long are playing mind games with other animals too. The flash of the Admiral's warning wings tells us that nothing is quite what it seems. If they confuse us humans too, then all the better for the Admirals of the future.

9.

Fire and Brimstone

Butterflies and the Imagination

The earliest writers about our butterflies were concerned with a matter to which few today give any thought. They wanted to know what butterflies were *for*. They must have some purpose, they reasoned, for otherwise God would never have created them. This was a problem because unlike, say, bees, butterflies had very little to offer to mankind. The Puritan physician Thomas Moffet had got hold of a medicinal recipe in which the 'venomous dung' of butterflies could be mixed with aniseed, hog's blood and goat's cheese to produce an effective cure.[1] But butterfly dung, venomous or not, is hard to come by. Lacking anything more useful to fall back on, men of learning were reduced to suggesting other reasons why God should have sent us the butterfly. Again, Moffet thought he had the answer. By being 'painted in colours more impressive than any robes', butterflies 'pulled down' sinful pride, whilst in the shortness of their lives they taught us to be mindful of our own failing condition. Butterflies both chided and warned. Everyone should be humble and ready for death, said Thomas Moffet's butterfly.[2]

By the time John Ray was writing his *History of Insects*, in the 1690s, it was possible to believe that God had given us butterflies purely for our delight. As Ray famously expressed it:

> You ask what is the use of butterflies? I reply to adorn the world and delight the eyes of men: to brighten the countryside like so many golden jewels. To contemplate their exquisite beauty and variety is to experience the truest pleasure. To gaze enquiringly at such elegance of colour and form devised by the ingenuity of nature and painted by her artist's pencil is to acknowledge and adore the imprint of the art of God.[3]

John Ray saw in the works of nature a reflection of the mind of God. To him butterflies were entirely benign, apart from the notorious cabbage whites. Some of his contemporaries were equally enthralled by the early stages of the butterfly's life cycle, the caterpillar and chrysalis, which were less beautiful but had other things to teach us. By mixing Christian philosophy with older, classical ideas of 'metamorphosis', they saw in the progression from lowly grub to 'angelic' butterfly a mirror of the human soul as it journeyed from birth to death and beyond it to resurrection. That, quite as much as their obvious beauty, is what first led people to take an interest in butterflies. The transformative power of that metaphor is just as strong in our own secular age. The lowly earth-bound grub and the celestial butterfly will go on feeding the imagination for as long as literature exists.

Naturalists today are less concerned with the mind of the Creator and more with what biological role butterflies might play. What difference would it make if every butterfly died out tomorrow? Quite possibly, very little. Yes, butterflies are pollinators but much less importantly than bees.[4] Even plants that seem designed for the tongues of butterflies, with their sugar syrup deep inside a long tube, depend more on day-flying moths, such as burnets, than butterflies. Presumably certain host-specific parasites would follow their butterfly victims to extinction. Possibly some crab spiders might go hungry. But, in Britain at least, no other form of life seems to depend on the survival of butterflies. The answer, in human-centred terms, is still the same as John Ray's: that butterflies are 'good' insects because they make us happy.

That is not to assert that butterflies are biologically useless. The adults, and still more their caterpillars and chrysalids, are food for birds, mice, lizards and frogs, as well as a vast number of predatory insects and spiders. Butterflies are neither more nor less 'necessary' than most other forms of life. They represent a way of life that evolved with flowering plants and has flourished for millions of years. Success, not usefulness, is the yardstick of evolution. You could sum it all up by saying that butterflies exist because they can.

People seem to have wondered about butterflies, and butterfly-like moths, since the days when we wore skins and lived in caves. There is, for example, a cave painting in the Pyrenees which was once thought to be of owls but which is a much better match for another nocturnal creature, the Eyed Hawkmoth.[5] Not far away in a second cave is another crude outline that resembles another insect with eyes in its wings: the

Peacock butterfly. Possibly it was the eyes that did it for these thoughtful cave artists: those unblinking orbs that stare out at the world with seeming hostility. Perhaps the men of the Stone Age incorporated these fearfully eyed insects into their philosophy of the world and its mysteries. We will probably never know.

Stylised butterflies reappear in certain artefacts of the Minoan civilisation in Crete around 4,000 years ago. The Minoans had evidently noted the similarity of a butterfly's shape to one of the key emblems of their culture, the double-bladed axe or *labrys*. The meaning of this axe is disputed; evidently it was a religious symbol of some kind, for images and remains of such axes have been unearthed in temple locations; they seem to be the approved representation of the thunder god or earth mother. But their effect was to draw butterflies into the belief systems of the Minoans. There was, by implication, something divine in the wings of a butterfly.[6]

The ancient Greeks took things a step further by explicitly equating butterflies with their idea of the human soul – so much so that butterflies and souls shared the same word: *psyche*.[7] Some of the images used to portray this relationship are disturbing. On one painted jar from the Attic period, around 600 BC, two naked black-painted figures are taking part in some obscene rite. The one on the left plays the reed pipes while drops of semen fall from his erect phallus . . . and promptly turn into a butterfly.[8] The philosopher Heraclitus held that the soul was an exhalation of fluid. Other philosophers, including Plato, believed that soul and semen were in some way connected, and the design on this antique pot seems to be making the same point. This belief was long-lasting, for the priapic man and the butterfly pop up again in stone-cut images from Roman times.

Butterflies also appear in the tombs of ancient Egypt, notably on a fragment of wall plaster now in the British Museum. It was formerly part of the tomb of one Nebamun, a 'counter of grain' who lived around 1400 BC.[9] It shows an everyday hunting scene in the Nile valley. Nebamun has left the counting of his grain to have a day out with his family, catching birds in the marshes. He is obviously having a good day for there are birds all around, including shrikes and egrets, three of which he has caught by the legs (while his cat has grabbed another by the wing). The kerfuffle has also disturbed a number of large reddish butterflies with white spots in the corners of their wings. They flutter above the lotus lilies and papyrus reeds which Nebamun's wife is collecting, perhaps to arrange in a vase back in Memphis. These are well-observed real butterflies, identifiable as the Plain Tiger, *Danaus chrysippus*, a migrant related to the Monarch which is still to be found in Africa and the Middle East today. Perhaps, by including them so prominently in his tomb chapel, the grain counter was hoping the butterflies would accompany him into the hereafter. He obviously liked butterflies. Perhaps, since human beings have always differentiated familiar kinds of animals, Nebamun even had a special name for his Plain Tigers.

We also find identifiable butterflies among the ruins of Pompeii. Among the mosaics that once adorned the floors of the wealthier houses preserved under the volcanic ash of AD 70 is one of a skull on top of fortune's wheel. *Memento mori* or 'remember you will die' is the merry message. Sandwiched between skull and wheel is a butterfly with apparently iridescent wings patterned with blobs of blue and half-moons of yellow and white. Again the artist appears to have drawn his inspiration from a real butterfly; it seems to be his best shot at reproducing the pattern and purple iridescence of the Lesser Purple Emperor, *Apatura ilia*.[10] Skulls and butterflies appear together in similar compositions from antiquity to the Renaissance. They symbolise death and the afterlife; the transformation from earthly grub to heavenly wings that we must all hope to attain when our time is up.

HOPE AND DAMNATION

Butterflies are fairly frequent ornaments on the illuminated manuscripts of the Middle Ages. Brown or white ones flutter in the foliage that surrounds the painted illuminations in the *Hastings Book of Hours*, a work of devotion made around 1480 for Lord Hastings (he who lost his head in Shakespeare's *Richard III*).[11]

More sinister butterflies appear in the famous triptych of Hieronymous Bosch, *The Garden of Delights*, painted around 1500. They are painted with great care and accuracy even though their natural bodies have been replaced by those of little demons. On the central panel a creature with the forewings of a Small Tortoiseshell sips nectar from a cardoon (a large blue thistle). All around it scores of sinners are having the time of their lives, making love, gobbling strawberries and riding in a cavalcade of fantastic animals. Bosch's precise meaning is not always clear but he seems to have chosen the butterfly, as well as his bright jays, hoopoes and woodpeckers, as symbols of the temptations of the flesh.[12] Bosch's sinners are attracted both to the beauty of their human partners and to all the fantastic animals, birds and flowers around them. But beyond the carnival of pleasure we can see, although they cannot, where it is all leading. Hell is just round the corner, on the third panel, awaiting their quailing souls. An alternative interpretation might be that Bosch's tortoiseshell butterfly offers the promise of salvation but that none of the sinners are interested because they are far too busy sinning. This is where it all leads, explains Bosch, with a sigh and a roll of his eyes, as he dips his brush into the red pot to paint another demon.

To make the point even more explicit, the beasts and birds of the 'garden' have ghastly counterpoints in the hell panel. Here resides our second butterfly, a demonic Meadow Brown. It seems to be officiating at a wedding – except that, by contrast with the diaphanous young women of the central panel, this bride is hideous. The sombre-coloured Meadow Brown seems to have been associated with the underworld. According to Franz Schrank, a German entomologist writing in 1801, it was the child of 'dusky Proserpina', the queen of Hades.[13] In consigning the butterfly to the genus *Maniola*, meaning 'little ghost', Schrank may also have been playing on the word 'mania', a mad frenzy of the sort that certainly animates *The Garden of Earthly Delights*. Bosch's Meadow Brown is a devil from hell, and as it turns its hideous beaked head towards the cowering bridegroom-soul you can almost hear it saying, 'I told you so.'

According to the story told by the Christian fathers, the rebel angels lost their bright wings when they were evicted from heaven and thereafter took on an ugly form emblematic of their wickedness. Pieter Bruegel the Elder (c.1525–69) captures that moment in his painting *Fall of the Rebel Angels*.

High up among the clouds, the host of the wicked is being beaten back by a swarm of angels led by St Michael (distinguished by his red cross). The rebels are turning into various slithering or flapping forms in front of our eyes, some with the wings of insects, others more like bats. But their leader, Lucifer, still wears his original, bright wings.[14] They are those of a European Swallowtail, although the artist has exaggerated the length of the tails perhaps for sinister effect. Bruegel may be reminding us of the beauty Lucifer once had, back in his glory days before the fall (Lucifer means 'the bringer of light'). In a moment he will lose his bright wings and turn into the horned, bat-like creature we know so well.

The polymathic German artist and engraver Albrecht Dürer (1471–1528) could draw plants and animals in extraordinarily realistic detail, but also, like Bruegel, he could include insects in his pictures to make a moral point. In his large canvas *The Adoration of the Magi*, it is easy to miss the two butterflies fluttering close to the skirts of the Virgin Mary. The pale one is a rather faded Painted Lady, a species Dürer might have known as the 'Belladonna'. The other is that 'gilded butterfly', the Clouded Yellow; and, whether or not by coincidence, this butterfly is a male. The artist's positioning suggests that their purpose is to echo the Virgin and Child, and thereby underline the message that lies at the heart of the picture. Dürer would have known his Greek philosophy, which held that each human body contained a soul symbolised by the butterfly. But the butterfly was also a well-known metaphor for the transience of life and the inevitability of death. Earthly existence is as brief as the life of a butterfly; only the soul is immortal. In allowing butterflies into his picture, Dürer is reminding us that although the Adoration was but an instant in time, the glory of God lasts for ever. The image is both transient and timeless.[15]

There is a common thread running through these emblematic butterflies from ancient times to the European Renaissance. Indeed, the thread runs on into early modern times as we have already noted in the case of the Red Admiral. Throughout the Dutch Golden Age, painted butterflies act as coded references to Christian beliefs about life and death, salvation and damnation. The butterfly might symbolise goodness or evil, depending on the species and its role in the composition, but either way they were brought in to add depth and meaning.

It seems, then, as if artists have long seen something spiritual in the butterfly. They are set apart from other winged life by their colours and graceful flight; they are the closest thing the natural world offers to our idea of the spirit. To a variety of cultures from ancient times to the near-present, butterflies have represented the visible part of the human soul.

PSYCHE AND HER ASSISTANTS

I am uncomfortable about the idea of a soul. When I was a child they told me not only that I had one but that it was the most precious part of me. When I died my body would decay but my soul would be whisked off either to heaven or to hell depending on how things went. I did not like the sound of that. Even as a child I would much rather have been properly dead than hang about indefinitely as a disembodied wraith. But, as I discovered, there were other possible options. Some said it was more likely that my soul would be recycled into some other living creature, a nice girl's pet guinea pig, if I was good, or, if not, maybe a starved, beaten donkey. Or an earwig, or the lowly worm. Buddhism seemed to apportion rewards and punishments in a similar way to Christianity. After a while, I began to doubt whether my soul, if it existed, was of any interest to divine beings of any hue, assuming *they* existed. The concept of a soul seemed to make no sense without God; ergo, if God did not exist then neither did the soul. Unless of course you believed in the supernatural, and you stopped doing that the day you caught Dad instead of Santa tiptoeing in with the Christmas presents and stumbling on the mat.

For today's earthly sceptic, the soul is a metaphor, a convenient word for our deepest and most intense feelings – love, desire, doubt, suffering. But even on that reduced definition, you constantly meet people who seem to manage all right without one. No one alive has ever seen a soul and so no one can say with any confidence what it looks like. This was a problem for medieval artists. Since they could not depict the genuine thing they had to find some embodiment, some visual metaphor. The obvious place to look for a soul-symbol was the natural world.

For the equally unimaginable Holy Spirit, the Bible offered artists a suitable metaphor in the dove. Perhaps this choice had something to do with the bird's soft expression, or its consoling cooing, or the possession of wings that resemble those of imagined angels. As Bible readers will know, at the moment Jesus was baptised in the River Jordan the Spirit of God descended from the parted clouds in the form of a dove, surrounded, according to St Matthew, by playful flickers of lightning. Another dove returned to Noah in the Ark carrying in its beak an olive branch to inform the patriarch that the flood waters were subsiding and that it was nearly time to let the animals out. Both dove and olive branch are emblems of peace.

Visualising the soul offers a tougher challenge because it is rooted within us, not left to flutter outside like a bird. Soul is an expression of feeling, a thing of the mind, close to the idea of a dream. Psychology takes its name from the Greek word for 'soul', *psyche*. According to Freud, the psyche is ever present in the unconscious, to be released in dreams when deep feelings well up like bubbles from mud at the bottom of a pond. Psychologists are to dreams what medicine is to the body, but not even Freud could inform us what the psyche looked like. The Bible is no help either, and neither is Socrates who seems to have invented the idea of a non-physical presence lurking within us. It was Aristotle, the 'natural philosopher', who first made an explicit connection between the soul and the butterfly.[16] He noted how the butterfly progresses from condensed dew – his explanation for their origin – to the greedy crawling caterpillar, then through an apparently 'dead' coffin stage, the pupa (*nekydallos* or 'little corpse'), and on to the climactic moment when the creature creeps from its 'tomb', unfurls its brilliant wings and takes to the skies. Aristotle saw in this transformation a revelation of how the sensitive soul (*anima* in Latin) propels the insect along its complicated journey from worm to the perfect *imago*. So it was, he taught, with all life, including the lives of human beings. Unlike butterflies, we are blessed with reason, but, in much the same way as the butterfly's journey from crawling worm to winged flight, so the human soul journeys towards perfection in its eventual release from the body. Hence we live our earthly lives in the sinful caterpillar state and it is only after the death of the body that our soul finally escapes into the spirit world.

This immortal soul was often represented by a butterfly. One such butterfly sits on the tomb of Beethoven in Vienna as a symbol of the immortality of great art. When classically trained artists wished to create an image of Psyche, they gave her the form of a beautiful young woman with a pair of insect wings – the round, veined wings of a butterfly or moth. In some adaptations she is shown wingless but with a white butterfly hovering suggestively nearby. As the artist saw her, Psyche was a sort of double being, a combination of a flawless female body and a perfect white butterfly. And even when rational people ceased to believe in this stuff, artists such as Canova continued to sculpt her image. For, quite apart from anything else, the naked woman-butterfly was lovely to look at.

AS MOTHS TO A FLAME . . .

Even back in the 1590s, long before science had banished the obscurities of superstition, Thomas Moffet could scoff at the 'silly people' who insisted that 'the souls of the dead did fly in the night seeking light'.[17] Moffet's religion told him to frown on non-biblical explanations, and the idea of ghosts hanging about in the darkness seemed to him to be a joke. So it does to us, at least until some moment of crisis or disaster, such as the sudden death of a loved one. Then, at least for a short while, we feel the need to look beyond the rational and hope there is some existence beyond earthly life after all.

Religious people endeavour to search beyond themselves to make sense of life, to look for consolation or redemption, towards some external, heavenly power for hope or forgiveness. It is human to want to feel that death is not final; that something happens beyond it, on the far side, even if it is not necessarily heaven or hell. A swarm of migrant butterflies over the Western Front during the First World War got many soldiers thinking of dead comrades and passing souls; there was an added poignancy to the fact that the butterflies were unaffected by the war, their beauty still unblemished.[18] Bereaved mothers have sometimes seen in a visiting moth the soul of their recently dead child. Moths, after all, live in the darkness but seek the light.

Many poets and writers from ancient times to the present day have alluded to the *anima* that seems to dwell inside a butterfly or moth. None did so more hauntingly than Virginia Woolf who saw in a dying moth a glimpse of this 'divine energy' at work in a small, insignificant body, 'as if someone had taken a tiny bead of pure life and decking it as lightly as possible with down and feathers, had set it dancing and zig-zagging to show us the true nature of life'.[19] It seemed to her that some external force was at work in its flutterings at the window followed by a sudden surrender to death. Was this what Aristotle saw too? In Virginia's case, the moth seems to have been part of a train of thought that ended in the river a few weeks later, when she drowned herself. Her last book, *The Death of the Moth and Other Essays*, was published posthumously.

Moths baffle us with their strange habits. Butterflies we can watch on their daily rounds, feeding from flowers, basking on warm stones, even going to bed in the evening when some of them roost communally and turn the grass stalks into a butterfly dormitory. We relate to butterflies because we too enjoy drinking, basking and sleeping. But the lives of moths, being creatures of the night, are hidden from us except at odd moments when we find them, huddled and inert, on the window pane, or under the porch light. Moths risk death – indeed, they seem to welcome it – in their helpless attraction to the flames of candles. We have forgotten now what every home once knew: the smell of singed moth. And not just the smell. W. H. Hudson described the sound of a burning moth as 'indistinct, faint [like] a dream in the night'.[20] Round and round the burning candle they go, like the moons around Jupiter (suggested Nabokov), drawing closer and closer to their fiery destiny until, at length getting stuck in the hot candle grease, they achieve their apparent desire and frazzle. Of course we have the advantage of knowing that the moth 'wants' nothing of the sort. Artificial light simply distorts its ability to navigate using natural beacons such as the moon, and so the insect is drawn to its doom as helplessly and inevitably as a spider going down the plughole.[21] But it is the nature of folklore to draw the wrong conclusions from correctly observed behaviour and to assume that the moth is committing suicide. Some compared the flaming insect to a soul drawn heavenward by the divine light. Others saw in it the trials awaiting the Christian on his journey to God: Joan of Arc in the flames, perhaps (and before leaving her there we might recall the legend that swarms of white butterflies accompanied her standard into battle). Don Marquis, in his guise as a cockroach narrator, claimed to have heard from the moth itself that 'It is better to be happy for a moment/and be burned up with beauty/ than to live for a long time . . .'[22]

Miriam Rothschild greatly admired a painting by Balthus titled *La Phalène* (The Moth). At first it is hard to spot the moth in this very odd picture. What we see instead is a pubescent young girl standing naked in her bedroom, reaching out towards the oil lamp. And then we notice it, a supernatural moth, more spirit than insect, glittering in the light of the lamp.

Meanwhile, completely ignored, a second, less conspicuous, far more realistic moth is quietly crawling up the bedclothes.[23] Balthus never explained his paintings but he seems to be suggesting that there are two ways of looking at things: the dream and the reality. For philosophers and poets, and sometimes artists too, it was the dream that had the greater emotional impact. Psychic moths and surreal butterflies excite our imaginings; they draw forth thoughts from the depths. Real moths are much more easily ignored. They can be left to the entomologist and the illustrator.

BUTTERFLY DREAMING

Perhaps it was a sign of growing scepticism when the British ceased to see moths and butterflies as souls and turned them into fairies. The Victorians liked fairies, not because they necessarily believed in them, but because, like science-fiction addicts in the twentieth century, they enjoyed fantasy worlds. Victorian fairies took human form; in their enchanted, night-time world, they glowed faintly, with an incandescent light. Some artists showed fairies with brightly coloured wings, often copied from a real butterfly or moth: the eyed wings of the Emperor Moth were a favourite. Actual butterflies sometimes joined the fun as winged fairy-horses, pulling an acorn-sized chariot through the air. The artist Joseph Noel Paton was inspired by *A Midsummer Night's Dream* to paint two huge canvases that swarm with fairies and more-or-less realistic bugs all so mixed up that it is hard to tell which is which.[24] One of the Fairy Queen's attendants has the wings of a Swallowtail, recalling Bruegel's rebel angel except that this angel is comely and would indeed grace a centrefold in *Playboy*. The Fairy King wears a butterfly hat. Of course it is all a dream and so they are excused the earthly limitations of science.

I used to dream about butterflies. They are my own proof that we dream in colour. My night-time visitors were intensely coloured. The Red Admirals flickered with fiery red; the exotic *Graphium* swallowtails were as jungle-green and light-catching as they are in life, and the shine of the morphos, those great blue butterflies named after the god of sleep, winked on and off as they do under the South American sun. I think my butterfly dreams were about longing. I would stand beneath a great flowering tree, a veritable Tree of Life, half-blinded by its scent, watching the clouds of butterflies drawn to the blossom. Mostly, I remember, they were familiar species but in unlikely combinations – emperors, swallowtails and fritillaries all sharing the same flowers. But sometimes – especially in deep, smothered, mothy dreams – the birdwings would come, with even larger, wildly iridescent forms that never existed outside my sleeping imagination. The bush became a Christmas tree of living lights. But if these were dreams of desire, what was it that I desired? To catch them, possibly. At least I remember sometimes leaving that phantasmagoric tree for a moment to rush home and fetch my kite net. It might have rounded off things nicely to bag those dream-butterflies and set them all in a row, but they were always just out of reach, or maybe there was a hole in the net. Mostly I just watched them and was glad.

I am no psychologist and my interpretation of dreams is plain and straightforward. I loved butterflies and at one time my love for them took the form of wanting to possess them. They were not, in my view, metaphors for anything else, nor disguises for wishes and desires unconnected with butterflies.

But, awaking from such dreams, I did wonder what I, like anyone else who has ever collected butterflies, was really seeking. Did we kid ourselves that it was in pursuit of a legitimate scientific objective? Was it a childhood fantasy that would not let go? Or was it at bottom some feral, atavistic instinct of the same kind that makes us want to shoot animals or catch fish – or even twitch birds? Was it, in another sense, about chasing a dream? We all want places where we can be ourselves. It might be, for those of a more gregarious temperament, a football match or a club dance. For me, in my mixed-up teens, it was to be out there among the butterflies. I had dream-butterflies fluttering around inside my head even when I was out in some wood or hillside stalking real ones. Even now there are encounters with butterflies that I seem to remember but which were almost certainly only dreams. And possibly the other way round. When I first encountered a great morpho butterfly in languid flight along the forest path in Costa Rica, flashing like a police car's blue lamp with every trip of its wings, I had the distinct sensation of dreaming. It was the same when, watching a Purple Emperor basking with outstretched wings, the sun suddenly ignited its blinding iridescence, I felt the urge to pinch myself. But the effect depends on being alone with your thoughts. In company, reality reasserts itself immediately. Friends are there to stop you dreaming.

THE INNER BUTTERFLY

Miriam Rothschild once combed the anthologies for words that had been used to describe butterflies. This was her list:

Simple
Gilded
Angelic
Joyous
Careless
Idle
Dizzy
Chaste
Languid
Silly
Peerless
Elegant[25]

To which I would add 'gentle' and 'worthless'.
Among butterfly metaphors we find:

Grace
Elegance
The soul
Naturalness
Freedom
Happiness
Purity
Transformation
Vulnerability
Resurrection
Hope

I knew a distinguished nature conservationist who let it be
known that he was fond of frogs. Word got about and whenever
anyone wished to give him a present it was invariably something
appropriately amphibious. Soon the poor man could hardly

escape the glare of plaster frogs from every shelf; his house was stuffed full of froggery. By the same token my parents' home ended up with butterfly images everywhere, on biscuit tins and jars, on wall calendars and knick-knacks. My birthday card invariably incorporated a butterfly; they thought it would please me. There was even a case of real butterflies, chosen for their 'autumn tints', above the telephone in the hallway which reminded me of Philip Larkin's line, 'The case of butterflies so rich it looks / As if all summer settled there and died.'[26]

I once made a list of all the butterfly images I could find in a supermarket, on organic food, on the healthcare shelves and, appropriately enough, on slabs of butter. A butterfly has a unique shape which we recognise instantly; a simple silhouette is enough. We associate them with sunshine and nature and all that they imply about wholesomeness and health. One of the things which butterflies help to advertise is naturalness. The Marbled White on the shampoo bottle, for instance, hints that the product contains more natural ingredients and fewer chemical additives than its rivals, and no doubt leaves your hair smelling fresh and outdoorsy too. We do not need to know anything about real Marbled Whites to get the message.

Another butterfly – a Monarch – was used to advertise the Open University. On one of the posters an intelligent-looking, middle-aged woman is thinking hard. She is presumably wondering whether she should, as the advert puts it, 'join the Open University and change her life'. The butterfly fluttering nearby symbolises knowledge, obviously enough, but it also suggests transformation and the freedom which knowledge brings. The best images of this kind are the ones that create the desired response in the consumer immediately without needing to spell it out. Advertisers are traffickers in human souls; they are canny observers of social trends. Their research tells them what people want and they hire artists to create appropriate metaphors of desire. The exact meaning of these billboard butterflies can be left ambiguous. It might be freedom or joy or nature or anything else loosely associated with butterflies, but the message is always positive. Butterflies sell.

Of course real butterflies are not 'free' in any meaningful sense. They lack free will and are prisoners of their instincts and of their genes. No real-life butterfly is happy or sad, nor is it 'chaste' or 'hopeful', nor 'idle' nor 'angelic'. Butterflies are not even particularly 'fragile'; they can absorb a lot of damage, as the faded and tattered survivors at the end of the season remind us. But a butterfly does have wings and it can fly. Many of us must have longed, sometimes, for a similar set of wings to float away from a trapped existence, like Ria, the Wendy Craig character in the seventies sitcom called – for that reason – *Butterflies*. Van Gogh must have had the same idea when he painted *The Prison Courtyard* in which convicts tramp in a circle within the high, oppressive walls of a prison yard. What tugs the heartstrings most are the tiny white butterflies fluttering high above their stooped, grey-capped heads. Are they a symbol of freedom or do they represent life beyond this living death?

Butterflies also stand in some vague but powerful way for self-realisation. Everyone likes to be healthy. A few years ago, Natural England, the government's nature conservation body, was earnestly peddling the notion of nature as a kind of 'well-being' spa – a 'green gym' – the idea being that the closer your contact with nature, the healthier you will be (Natural England's medical advisor, rather happily, was a Dr Bird). Others promote the New Age idea that you can learn more about yourself through nature. *Learn* from the butterfly, I read, on one self-help website. If only you stop to listen, butterflies have the power to teach you about your inner feelings and so make the right choices in life. For example, by watching the Small Tortoiseshells and Peacocks in the garden you can compare your life with theirs and ask yourself hard questions. Are you are as lively as they are? Are you equally happy and carefree, or as fulfilled? In this representation, the flight of butterflies seems in some mystical way interconnected with our own *psyche*. It informs us that change does not have to be 'traumatic'. A further way in which the New Age butterfly can help is in therapeutic exercise. You could, for example, try wrapping yourself in a blanket to make your own 'cocoon', then slowly release the folds and emerge like a butterfly. Feel better now?

Well, if it works for you, then it works. Faith, they say, is nine parts of the cure. Alternatively you might agree with Mark Twain that 'when we remember that we are all mad, mysteries disappear and life stands explained'.[27]

Perhaps butterflies still stand in some way for the soul. We have become consumers and self-believers, but we also seek something beyond ordinary existence. We are, despite everything, still followers of the metaphorical butterfly.

10.

Silver Washes and Pearl Borders
Painting Butterflies

Until the camera came along to take photographs of living butterflies, the standard image of a butterfly was a grossly distorted one. It shows them not as they are in life but in death, mounted on a pin in a museum. The wings are held out flat at ninety degrees to the body – a posture which would sorely test the ligaments of any butterfly. It is nevertheless the shape which we instantly recognise as 'butterfly', whether in bow-ties or the pasta the Italians call *farfalle* (after *farfalla*, a butterfly). Yet in life butterflies open their wings flat only to bask in the sun, and some species never do it at all, but settle with their wings tight shut. Even when the wings are wide open, they always slope downwards more than mounted specimens, with the hindwings much closer to the butterfly's body.

Field-guide butterflies are shown that way for two main reasons. First, they were copied not from live butterflies but from museum specimens which cannot move or fly away. Moreover, the illustrator's main job is to facilitate identification. This is best achieved when all the butterflies are shown in the same way. This is the essential difference between illustration and art. The job of the illustrator is to present standardised images which can be compared with one another. The artist attempts to show the butterfly as it is in life – or something of its essence – and as the human eye perceives it.

Before the twentieth century, many artists displayed butterflies in settings that recall the careful but artificial arrangements of still-life paintings. The subjects are often beautifully and accurately painted but there is also something unreal about the way they are shown suspended in mid-air. Without camera images or slow-motion films to guide them, artists had no clear idea of how butterflies actually fly (they do so by pulses of the wings in three dimensions). Hence, most of these painted insects look like what they are – a dead butterfly minus the pin. Even with the greatest artists, such as John Curtis and Moses Harris, the butterfly seems more like an icon than a living, moving insect; you sense that they are not really alive. They are artist's dummies, objects in a composition, though very beautiful ones.

One of the first to paint butterflies accurately, and for their own sake rather than as decorations or symbols, was Joris (or Georg) Hoefnagel (1542–1601), a travelling artist from Brabant in what is now the Netherlands. Like most jobbing artists, he did what he was paid to do. He was famous for his miniature work, akin to medieval illuminations – Hoefnagel has been called the last illuminator – and they include well-observed, faithfully drawn images of animals, plants and, especially, insects. His most important commission was from the immensely rich Emperor Rudolf II of Austria to paint curiosities from the Emperor's *Kunstkammer* or art room of treasures. Hoefnagel set to work and, since strict accuracy was the Emperor's wish, he used all his skill to produce images of preserved insects that seem to have come back to life. His trick was to paint shadows beneath each specimen to create a three-dimensional *trompe l'œil* effect in which these dried-up cockroaches and beetles seem about to walk off the page. He painted butterflies too, including a realistic Heath Fritillary clinging to a frame while a winged Cupid whispers something in the ear of a naked woman who is probably meant to be Psyche. This is a butterfly that is at once real and mythic – an actual fritillary that also embodies a spirit. Hoefnagel was a one-off, though his son Jacob, who followed him into the trade, used a magnifying lens for the first time. But science, as it is understood today, was not their aim. Rather Jacob recycled his father's *trompe l'œil* insects and other tricks of his trade into a kind of manual, a 'pattern or copy-book for artists'.[1]

England usually lagged well behind European fashion and it took a century or more before English artists began to add butterflies to their portfolio. The stimulus was provided by a Dutch artist, Maria Sibylla Merian (1647–1717), daughter of a map-maker, who grew up among the copper plates and tools of an engraver's workshop. Merian lived in Frankfurt until 1685, when, newly divorced, she moved to the Netherlands with her two daughters to join a community of 'primitive' Protestants. There she earned a respectable living, and a growing reputation, as a painter of flowers and insects as well as teaching, and making and selling artist's pigments. Her first published collection of paintings, her *Blumenbuch* (or 'A first book of flowers'), was followed by the much more outré *Raupenbuch*, about caterpillars 'and their remarkable diet of flowers'. Caterpillars, unlike adult butterflies, are difficult to preserve and so Merian painted them from life. Her most famous work, completed when she was nearing 60, was her *Metamorphosis or Transformations of the Insects of Surinam* (Dutch Guiana): sixty luscious arrangements of plants, animals and insects from tropical South America. Merian had taken advantage of trade connections among her co-religionists to set sail for the remote colony in 1699. As she recalled later, 'Everyone is amazed that I survived at all. Most people die there of the heat . . .'[2] Her watercolours, subsequently engraved and hand-coloured for publication, were something new, representing a close engagement with living insects as opposed to dead ones in a cabinet. Their arrangement recalls the formalised flower paintings of the Dutch masters but Merian's insects are not 'still life'. Shown against a plain background they flutter and buzz or perch on a nibbled leaf or bunch of fruit, while caterpillars slither and crawl over the vegetation. Merian loved natural patterns and shapes, especially spirals and curls. The antennae of her moths and beetles twist and coil as do the tendrils of her vines and passion flowers, the tails of her lizards and the roots of her sweet potatoes. The demand for copies of Merian's *Metamorphosis* went far beyond learned circles in Amsterdam. One was purchased by Sir Hans Sloane; another went to the Royal Collection in Windsor. Her work inspired the first generation of insect artists in England, such as James Petiver, and in doing so provided the artistic template for the luxurious butterfly books of the eighteenth century and beyond, a mixture of the natural and the artificial, all arranged in a tasteful, harmonious composition.

Merian's style and close study of insect life cycles inspired a German artist called Weiss who, crossing over from Hanover to London, changed his name to Eleazar Albin (from Albion, perhaps, or maybe a pun on Weiss/white/albino?).[3] Albin had a wife and at least three children to support and, to judge by the engraving of him in his *Natural History of Spiders*, on horseback and wearing his finest clothes, he had social ambitions. Albin began to paint in Merian's style, rearing butterflies and moths to paint and incorporating all the life stages into his compositions. He even included the parasitic wasps and flies that sometimes emerged from the chrysalis instead of the expected butterfly.

The process of turning the artist's compositions into engraved plates was slow, laborious and expensive. The artist's original watercolour had to be copied by the engraver on to a copper plate for inking and printing. Copper is a soft metal and only a limited number of prints could be run off before the plate became worn and unusable. The engraving would, of course, be only an outline, although patterns and shadow could be reproduced by hatching. For coloured copies of the work there was no resort except hand-colouring, for which the artist normally employed assistants. Albin's daughter, Elizabeth, became an adept at this secondary art form. To offset his investment in time, paints, paper and plates, the artist needed to find subscribers. Fortunately there was now a ready demand for such pictures from the well-to-do, even for a set of plates like Albin's *Natural History of English Insects* costing thirty shillings uncoloured, three guineas coloured, 'half paid down and the other half on delivery'. His subscribers included members of the peerage and Fellows of the Royal Society, even royalty. Some of them offset the artist's expenses further by allowing their names to appear as sponsor for a particular plate. The name of John Philip Brayne FRS, for example, is associated with a rugged design of the Large Tortoiseshell including a well-nibbled spray of elm leaves. The elderly Sarah Bodville, Countess of Radnor, was partnered with Albin's arrangement of the life cycles of the Brimstone and the Black-veined White – a particularly attractive composition. But when she died Albin shrewdly scrubbed out her name and substituted another.

Albin's pictures are creditable for their time but much better work was to follow later in the century by gifted artists such as Benjamin Wilkes and Moses Harris. What is striking about all these productions from Georgian England is that they are much grander, and consequently more expensive, than they strictly needed to be. They were, you might say, art-driven rather than science-serving. Wilkes, who illustrated two gorgeous butterfly books in the 1740s, was at least as concerned in composition and making attractive butterfly patterns as in furthering knowledge; indeed his ghost writer candidly considered him, 'for want of learning quite incapable of writing a book'.[4] Nonetheless Wilkes's butterflies have much more bounce and energy than Albin's; his fruits are plump and peachy, his flowers crisp and luscious. And a hand-coloured copy of his *English Moths and Butterflies* cost nine pounds, the equivalent of three months' wages for an agricultural worker or domestic servant.

Arguably the finest of the Georgian picture books was *The Aurelian* by Moses Harris, first published in 1766.[5] Its justly famous frontispiece is an engraving of Harris himself, reclining in a woodland setting with his insect trophies

displayed in their oval box and a fellow collector (or maybe
Harris again) stalking the lane nearby. Each folio plate is
artfully arranged, usually with some artistic adornment such
as a vase of flowers or, in the case of the Painted Lady, pieces
of broken pottery, an old clay pipe and a discarded mussel
shell of the sort used by artists as a palette. (The reason for
the broken pots may not be obvious: the Painted Lady feeds
and lays on thistles, and thistles commonly grew on rubbish
dumps.) The plate showing the Death's-head Hawkmoth is
dedicated to Harris's 'ingenious Friend and Benefactor' Dru
Drury, perhaps as a mordant private joke. *The Aurelian* has
a unity of composition which suggests that Harris, like Albin,
bred each species and studied their natural postures. The
challenge for all these artists was to fuse technical accuracy
with the demands of art, to produce something that was at
once lifelike and also satisfied the canons of taste. The
butterfly, you might say, was where art and science met on
equal terms.

Moses Harris's passion for butterflies led to the invention
of his 'colour wheel', designed to assist the artist with the
interplay of colours found in the wings of butterflies. His
booklet, *The Natural System of Colours*, dedicated to Joshua
Reynolds, was the first in English to explain how the most
subtle of shades (his word was 'teints') can be created out of
just three 'primitive' colours, red, yellow and blue.[6] By the
same means he demonstrated what is nowadays called the
subtractive mixing of colours in which the primary colours
can be superimposed to form black. Harris stands high among
bygone entomologists: he drew and engraved some of the best
butterfly images of his time and for long afterwards; he seems

to have invented some of the more poetic names of our butter-flies and, especially, the larger moths; he revived the old Society of Aurelians and he produced the first pocketbook for ento-mologists, one that incorporated for the first time the binomial 'Latin names' of Linnaeus. And, by studying the natural colours of butterflies, he also came upon a new way of creating and mixing colours.

The last artist to paint butterflies in the same spirit was Henry Noel Humphreys (1810–79).[7] His watercolours bore the same pretty arrangements of butterflies and flowers, but his flowers were no longer in vases and there were no more clay pipes or mussel shells introduced purely for decoration. Everything in Humphreys's pictures was there for a definite purpose: to inform. In the book he illustrated for John Obadiah Westwood, *British Butterflies and their Transformations* of 1841, the plates were printed lithographically before being hand-coloured and so were a closer match with the original water-colour than the earlier copper engravings. It also meant they were much cheaper.

Humphreys, like his forebears, made his living from art. Though a keen observer of butterflies and moths, his primary interests were in medieval illumination and archaeology (he was an authority on ancient coins). His contemporary, John Curtis (1761–1862) was the exact opposite, a specialist who painted and engraved insects, and nothing but insects, for upwards of forty years.[8] He was the first to depict them in a fully scientific way, aiming at absolute accuracy in every tiny detail. (He once reproached another's drawing by pointing out that he had missed a bristle; did he not know that this species has *thirteen bristles*, not twelve?) In that sense, Curtis was the first full-time insect illustrator and almost all of his enormous output was destined for the same source, the journal *British Entomology*. He was the entomological equivalent of his better-known namesake (but no relative) William Curtis, the illustrator of *Botanical Magazine*. It was John Curtis, and not Moses Harris or Henry Noel Humphreys, who represented the future for insect illustration. The gap between art and science was now too wide to bridge. For better or worse, butterfly illustration had become a handmaid to the study of entomology.

When I was about 12, my father found an old book in a second-hand shop and bought it for me. It was *The Complete Book of British Butterflies* by F. W. Frohawk (1861–1946): a neat, serious-looking volume bound in plain green buckram. For years it was my most treasured book, the only one to describe and illustrate the life stages of every butterfly in full colour, along with all the most significant varieties or 'aberrations'.[9] It was Frohawk who put the seal on painting butterflies as they are displayed in collections, set with mathematical precision. But he also drew butterflies from life, neat little sketches reproduced in halftone, a butterfly laying eggs, or asleep on a twig, or settling down for the night among ivy leaves or long grass. He even noted which direction the wind was coming from as he sketched. As far as I know, Frohawk's drawings are the first to show butterflies live in the field. In that, as in much else, he was the fore-runner of modern butterfly artists.

Frohawk had a well-rounded talent. Not only was he one of the best illustrators of his day but his knack of rearing butterflies has rarely been equalled. He was the first to breed every British species from egg to adult, even the Large Blue

which required some complicated experimental work involving
a nest of ants and a walnut shell. He was, said Norman Riley,
curator of butterflies at the Natural History Museum, 'the hub
around which amateur lepidopterists gravitated'.[10]

Perhaps it was because of his unusual name that I imagined
Frohawk to be rather grand and elegant of person, looking,
perhaps, a bit like Sherlock Holmes. In fact Frederick William
Frohawk was a stocky man with a big, square head. He wore
thick granny specs and was nearly blind in one eye. Out in
the field with his net and sketchbook he wore a tweed jacket
of his own design, with extra pockets, a waistcoat, matching
woollen breeches and thick socks or gaiters wound around his
calves: heavy attire for a warm summer's day. He sometimes
carried a catapult in one of his pockets, describing it as 'a most
useful weapon for collecting small birds'.[11] Nor was he as rich
and grand as I imagined. Like most specialist illustrators, he
scraped a bare living, mainly from work for the *Field* or the
Natural History Museum. He also drew an enormous number
of illustrations for the *Entomologist* but those were all unpaid.
His first wife died young, leaving him with two daughters to
look after, and in order to buy a house for his second wife and
a third daughter he was forced to sellhis butterfly collection
to his friend Lord Rothschild. That must have felt like losing
a piece of his soul even though it ensured that, in due course,
it would be preserved as part of the national collection.

Frohawk was incredibly unlucky. He published three
butterfly books, each a classic of natural-history publishing,
but none of them made him any money. The first, his magnum
opus, was the magisterial two-volume *Natural History of British
Butterflies*, a great book in every sense. Its folio-sized colour

plates are the first complete record of all the life stages of British butterflies. The work took him a quarter-century to complete and just when, at long last, in 1914, and at the age of fifty-three, he had completed the last painting, written the last line, and sent the great work off to his publishers, war was declared. The type press and plate blocks of the first volume had been set up and were ready for printing but wartime paper shortages made publication impossible. By the time the war was over people had other things on their minds: publication was delayed a further six years until 1924. To cover the costs of this expensive book, the publishers had insisted in obtaining sufficient advance orders. Once again Lord Rothschild stepped in to help the penniless author by buying the original paintings and drawings.

For his next, shorter and more affordable book (my one), *The Complete Book of British Butterflies*, Frohawk was obliged to paint a completely new set of plates in a smaller format. The hundreds of drawings and paintings it required must have taken him several more years. That book was published by Ward Lock in 1934 and it was followed, four years later, by his third and last book, also self-illustrated, *Varieties of British Butterflies*, which was about the rare and unusual forms that were such a draw for collectors. By then Frohawk was seventy-seven. Squinting over a lens had wrecked the sight of his remaining eye and he was thereupon forced to give up commissioned work. And, just as his first book had been a casualty of the First World War, so the second and third became victims of the Second. On the last Sunday night of 1940 the Luftwaffe firebombed the publishing heart of London. The entire stock of both books went up in flames, along with all the drawings, paintings and colour blocks. Hence the books could never be reprinted (nor, within his lifetime, was his projected book of British birds). 'This is a serious loss to me,' noted Frohawk in his calm way. 'No more royalties forthcoming and nothing to be done in the near future.' By the end of the war he was ailing and a year later he was dead. The wooden cross above his grave in Headley, Surrey, bears a worn carving of a Camberwell Beauty. He can stand, *par excellence*, for the strange spell butterflies have over a certain kind of Englishman.

In the past, full-time butterfly illustrators could find work in magazines, in museums and cigarette cards for tobacco companies. Today the main market is field guides leavened with occasional, relatively well-paid commissions for postage stamps or posters. A field guide might require a thousand images – perhaps two years' work at two or three paintings per day, week in, week out. Anyone who is not also an entomologist of rare dedication would soon give up. The colours of butterflies and moths are subtle, and many also have textures and shines that are hard to reproduce in paint. And not just the colours. Illustrators in the past rarely gave sufficient attention to the insect's body, tending to use one as a template for all. But butterfly bodies differ in detail from species to species, in their textures and bristles, subtly different colours and all kinds of almost microscopic detail in their legs and antennae. With the high standard of accuracy expected by today's public, there is little room for shortcuts. Many still prefer artwork to photographs since a skilled artist can subtly emphasise key features without distorting the likeness. No longer is he content to paint a flat outline, as before, but strives to create a more realistic impression by placing the light source at an angle to provide a slight modelling effect so that the veins seem to stand out a little from the surface of the wing. An insect's wings are rarely truly flat. Those of a dragonfly, for example, incorporate a kind of hydrofoil and if drawn flat they look merely diagrammatic. The artist must get the intricate network of veins right too, for a dragonfly expert would be quick to spot any mistakes. But imagine doing this, all day, nearly every day, for the rest of your life. The wonder is that anyone does it at all.

The Frohawk *de nos jours* is Richard Lewington. Now in his sixties, but looking at least a decade younger, he has painted insects, especially butterflies, moths and dragonflies, for most of his working life. Unlike some illustrators of the past, Lewington has never collected butterflies. Instead he bred them, and was inspired to paint all the stages of their life cycle by the subtle beauty of the chrysalis of the Red Admiral, which no artist had succeeded in capturing. Since then he has illustrated a great many field guides, including a very challenging one on micro-moths (the extremely numerous 'micro-lepidoptera') in which most species are smaller than a thumbnail. He draws and paints behind a big window overlooking the garden of his Oxfordshire home, his sketchpad on the tilted drawing board and his gouache paints, palettes and stereoscopic microscope at his right elbow, allowing him to peer down the lens before taking another brushstroke.[12] He prefers to work in natural light but, once the sun has gone down, resorts to a neutral-coloured lamp mounted on a bracket. He usually paints set specimens borrowed from a friend or a collection, but sometimes, as in the case of the micro-moths, from a live insect, sitting there quietly having its portrait done.

For the larger moths Lewington broke the mould by, for once, showing them in their natural postures instead of the way they look when pinned and set. And, no, he insisted when I asked the obvious question, he never gets bored – well, hardly ever. To him every micro-moth or bug or bee is quite distinct. He enjoys a challenge. For example, he worked out how to paint iridescent blue butterflies by building up colour washes so that they seem to reflect the light and shine as in life. He can capture the iridescence of a Purple Emperor in a way that seems to defy the limitations of paint. When I interviewed him for this book he was painting bees for a new field guide. And, yes, they are quite as difficult as butterflies, and, no, they are all quite different.

The limitation on butterfly illustration today is less on the fidelity of the artwork than on the standard of the printing. The ability of the printer to reproduce an illustrator's work exactly is surprisingly recent. Cheap colour printing using photographic methods was possible from the early 1900s but colour registration remained problematical well into recent times. Even Frohawk suffered from 'printer's fuzz'. Lewington is critical of the reproduction of some of his own work, pointing to examples where the colour is insufficiently saturated or incorrectly balanced. His happiest professional relationship was with Andrew Branson of British Wildlife Publishing when they generally managed to get it about right. That partnership culminated in the marvellous *Butterflies of Britain and Ireland* by Jeremy Thomas and Richard Lewington, first published in 2010 (though some of the paintings date from a previous incarnation of a book published by Dorling Kindersley). I think it is the greatest butterfly guide ever published; well-nigh the *perfect* butterfly book.

I doubt whether we will see its like again. Richard Lewington still paints butterflies in the way of his forebears. He has a better microscope than Frohawk, who would have owned one of those Victorian ones made of brass, but he uses the same kinds of pencils and paints and paper. The fine-art skills of Frohawk and Lewington are no longer fashionable and neither is the necessary anatomical knowledge. Their successors will, most probably, work on the computer to combine real and processed images into a simulacrum of reality – as Lewington himself did for his stamp designs for the Isle of Man. I think he will be a hard act to follow. Or, rather, he won't be followed.

A MOMENT IN SPACE

Painting a living butterfly in its natural setting, and as the human eye sees it, requires a special set of skills. Modern butterfly artists attempt to show us what the eye sees in the fleeting moment when we glimpse a butterfly, or as it might linger in the mind after it has gone; or, alternatively, a butterfly half-hidden in a tangle of vegetation; a blob of bright colour in the overall green shade. Success in painting butterflies as the eye sees them – not as the camera does – depends on matching the painted image with the one in the mind. Three artists in particular have sought to capture such brief encounters: David Measures, Gordon Beningfield and Richard Tratt.

David Measures (1937–2011) is often said to be the first artist to sketch real live butterflies (overlooking F. W. Frohawk who did so three generations earlier). Rather than aim at photographic realism, Measures tried to capture the gist of a butterfly in a series of rapid pencil sketches dabbed with watercolour. For him, the sketch *was* the picture. He would build up an impression of the butterfly and its habits through a succession of little drawings, showing it in flight from various angles or at rest with its wings closed. Measures worked out of doors with the minimum of equipment: a drawing pad or a sheet of paper clipped to a board, and a tiny box of paints. Sometimes he did not even bother with a brush but simply painted with his fingers using spit and fingernails for the detail. He finished off the sketches with pencilled notes. These were not necessarily things you would want to hang over the fireplace but they were, in their way, as true a record of the living butterfly as any in field guides. He sketched our last native Large Blues in drawings made the year before they went extinct; they look like animated blobs of watery ink and when you see a real-life Large Blue you realise how true to life his impressions are. They are images not of the butterfly per se but of what the eye *sees* of it.[13]

His friend Julian Spalding called David Measures 'the Audubon of butterflies'. His work first came to wider public attention in 1973 when he was featured in a programme titled 'David's Meadow' in the BBC series *Bellamy's Britain*. Three years later he published an anthology, *Bright Wings of Summer*, which was almost a nature diary, with each drawing timed and dated. Measures loved nature from boyhood and his work reflected the way he felt about it: 'There is a magnet in me,' he wrote, 'drawn to the subtle sense-aura of wild freedom, the porous exchange apparent in wild places and the richness of variety and subtlety which I miss inside a building.' That idea of 'porous exchange' is a telling one; he did his best to let nature into his art – and indeed to let nature guide his art. He loved the moment when a butterfly became 'reconciled to your presence, seems to allow a trust to exist, whereby both of you take part, each functioning in your own way, freely and co-existent'. So absorbed would he be, focusing his attention on capturing those fleeting moments, that he grew oblivious to passers-by and on at least one occasion was mistaken for a scarecrow.

At the same time as David Measures was developing his impressionistic style, another artist was painting more 'finished' pictures of living butterflies in watercolour. Gordon Beningfield (1936–98) was, like Measures, a keen naturalist from boyhood onwards but he had trained as an ecclesiastical artist specialising in glass engraving. He found an alternative career on television where he contributed to various country shows in the 1970s and '80s including the surprisingly popular *One Man and his Dog*. Beningfield was an enthusiastic participant in outdoor activities from dog handling to shooting and fishing. He looked the part, too, with his relaxed manner, cloth cap and sideburns, speaking with a soft Hertfordshire burr. Many remember his stint as a cycling artist along the byways of Dorset, in his tweed jacket and well-polished brogues, easel and paints strapped to his back, looking for views that reminded him of the novels of Thomas Hardy.[14] Beningfield became adept at painting soft, chocolate-box pictures of old England, its countryside, woods and villages. He even painted his own visual autobiography, *The Artist and his Work*.

Beningfield loved butterflies. In fact they were his first love. He had collected them as a boy and as a grown-up he decided to have a go at painting them. In 1978 some of these paintings were brought together and published as *Beningfield's Butterflies*. It became a surprise best-seller. We had long been used to butterflies as illustrations, displayed as in a museum. But Beningfield gave us the butterflies we see in the garden and in the countryside: a Red Admiral sucking the juices from a rotten apple; a Brimstone barely visible among the ivy leaves; a Comma among autumn foliage that counterpoints its wonderful raggedy wings. Beningfield's pictures reminded us how small butterflies are. He painted them as dabs and spurts of bright colour within their micro-world of grass blades, flowers and leaves. The paintings in *Beningfield's Butterflies* were exhibited on the day of publication and by the end of it every one of them had been sold; indeed they were so oversubscribed that the last ones had to be sold by drawing names from a hat. 'There's no market for butterfly pictures,' a well-known gallery owner had told him. But there was, at least for this kind of butterfly picture, pretty, accessible and at the same time true to nature. Though he is a little out of fashion today, Beningfield's were butterfly pictures you *would* want to hang on a wall.

The Post Office took notice and commissioned a set of butterfly stamps, an event more notable then that it would be now. Beningfield's stamp quartet, which went on sale in May 1981, was in the same style as his pictures, neat images in soft, green settings. The public loved them. I am less enthusiastic; the way the Peacock holds its wings is wrong and Beningfield's Small Tortoiseshell looks as big as a bat. They are pretty but unremarkable examples of miniature design. Richard Lewington, too, has painted butterflies for the Post Office, most notably for a set of ten stamps in July 2013. His images are as good as you would expect but, since the Post Office insisted on plain white backgrounds, they fly in a void with only the Queen's head, looking unconcernedly from the left, as company. Butterflies are hugely popular subjects for stamps but few of them really work. The format is too constricted. They looked better on Brooke Bond tea cards.

At length, Beningfield turned to other topics for he was far too much the polymath to anchor himself to one subject for long. His natural successor today is Richard Tratt (born 1953). Like Beningfield he paints butterflies more or less life-size within a much larger canvas. Indeed, Tratt's paintings are more like landscapes with butterflies than butterfly portraits. Unlike Beningfield and Measures, who worked in watercolour, he prefers to paint in oils. His butterflies, well-observed and accurately painted in thinned colours, feed, perch and fly within a more coarse-grained landscape. While his insects are usually finished off in the studio, his landscapes are often painted outdoors in order to capture the ever-changing effects of light and season on vegetation. Tratt, too, turned to butterflies not because they were necessarily the most marketable subjects but simply because he loves them. He has painted butterflies since his teens. He paints at all hours of the day, in sun, cloud and threatening storm. A favourite subject is the glow of the sun behind the Hampshire downs shortly after dawn with the butterflies waking up in the grass. He loves the scents and colours of high summer: 'Marjoram and wild thyme scent the air while scabious and knapweed cover the slopes in a perfect blend of colour.' 'Life doesn't get any better than this,' concludes Richard Tratt. 'Time to begin the next painting.'[15]

Butterfly paintings respond less to market forces than to the artist's preferences and emotions. The driving force for all the great illustrators, from the Georgian masters to F. W. Frohawk, and, in our own times, to David Measures and Richard Lewington, seems to have been not commerce but a deep love of butterflies. But increasingly, the joy people take from living butterflies is being matched by feelings of sadness, even desolation, when they are lost. The extinction in Britain of the Large Blue in 1980 made headline news, as the loss of any other kind of insect would not have done. Many people felt the force of that event and its dread finality. The bell is now tolling for many more butterflies. Progress and development have turned the joy previous generations felt in their presence to one tinged with apprehension. Fear for the future of the butterfly, but also for the barren world we are creating for ourselves.

11.

Endgame

The Large Blue and Other Dropouts

I left it too late to see the original, native English Large Blue. By the mid-seventies the sole remaining site, in one of the sheltered coombs of Dartmoor, was producing only a few dozen butterflies per summer. By 1978, after two summer droughts followed by a summer of rain, the number had fallen so low that it was felt necessary to enclose the colony with netting to ensure that the butterflies met one another. It didn't work. Only one last, poor season remained to them and after that there were no more Large Blues. The species was declared extinct in Britain in 1980. It was once found across southern England from Tintagel to Peterborough. Now it could not be found anywhere.

Yet the Large Blue lingered on for another year or two as a kind of ghost. A mystery man appeared on the news with a smile on his face and a twinkle in his eye. No, it wasn't really extinct, he insisted. *He* knew of a secret colony but no one else did and he would never reveal where it was. Did anyone believe him? In the context of lost birds such as the moa of New Zealand, Mark Cocker has suggested that we seem to have an inner need to snatch at straws 'to defer or circumvent the desolate finality of extinction'.[1] So perhaps a few people did.

Unlike other lost butterflies, this one met oblivion in a conservation-conscious age. Indeed, the Large Blue became more famous dead than alive. The loss of what was perhaps our most remarkable butterfly – the only one to adopt the lifestyle of a cuckoo and live as a parasite in an ant's nest – became a watershed moment in nature conservation. It marked the point at which the notion of reintroducing a lost species changed from an irrelevance to a serious proposition. The Large Blue had been desperately unlucky to encounter a series of poor summers when its numbers had fallen to their lowest. Desperately unlucky, too, in its timing, for had it survived a few more years we could probably have saved it. As it turned out, it was all a question of habitat. Jeremy Thomas had studied the butterfly's ecology and discovered, just too late, that what it needed most was short, sun-warmed turf with plenty of wild thyme and plenty of the right kind of ant. But most of its former sites were now overgrown with coarse grass and scrub, conditions that guaranteed disaster.

With the help of David Simcox, Thomas procured a licence to collect Large Blue eggs and larvae from southern Sweden and release them into what they hoped was suitable habitat in south-west England. These Scandinavian Large Blues seemed to accept their new surroundings and, slowly at first, they multiplied. Today the butterfly is well established once again in the Poldens and the Cotswolds with a few outliers elsewhere. So, you might say, we have made amends and the

Large Blue is British again. The project to reintroduce it became the forerunner of a series of ambitious 'reintroduction' programmes which have seen the return of the Red Kite, the European Crane and the White-tailed Eagle, and should, one day soon, ensure the return of the Beaver and the Lynx.

Summer 2012 was one of the most dismal in living memory. But if we human beings felt like taking to our beds for the duration, imagine what it must have been like for butterflies. When or where, in the steadily beating rain, would they find opportunities to emerge, to find a mate, lay their quota of eggs and so pass on their genes to the next generation? How, as the cold slowed their development, would enough caterpillars survive their predators and parasites to produce adult butterflies?

Yet somehow they did. Spring 2013 was also cold and wet but when the summer sun made a belated appearance towards the end of June, the butterflies responded. Some managed to put on a defiant show. I remember rubbing my eyes at the Dark Green Fritillaries and Marbled Whites thronging the thistle heads on the Hampshire downs and the White Admirals and Silver-washed Fritillaries crowding the brambles in the woods below. The dynamics of butterflies are to a large extent mysterious, but there must be some unseen system of checks and balances that gets them through the foul weather and into the fair. So long as there is sufficient habitat to sustain a population they seem to get by.

It was at the onset of that unexpected summer that I made a pilgrimage to find the reintroduced Large Blue. The most accessible place is Collard Hill on the Polden Hills near Street in Somerset. This place has its own website[2] and logbook, as well as a big friendly sign to welcome you to what you hope will be Large Blue heaven.

The day of my visit was warm and almost windless: perfect for butterflies, or so I thought. Yet I walked from one end of the hill to the other at the peak of the Large Blue season without seeing a single one. I began to think I must be too early – for this had been a peculiarly late summer – or maybe I was on the wrong hill. Spotting two elderly chaps bending down, evidently peering at something, I made haste to join them. 'Seen one?' I asked hopefully. 'Ah no, we think they might be taking a siesta.' It was, they told me, a bit too *hot* for Large Blues at the moment. Last year, of course, had been too cold. They had hardly appeared at all.

'Is there a temperature they actually like?' I asked. The man in spectacles and a straw hat considered my question politely. 'Well, they don't fly all the time. If you'd been here earlier you might have seen one. They keep their heads down at midday.' I said I'd heard that they don't fly in the early morning either, or not until the sun has warmed the ground. 'Well, if you were here around nine o'clock, you might have seen one or two.' He looked at his watch. It was nearly twelve noon. 'Why not come with me?' he suggested. 'We can work from the bottom of the slope to the top.'

My kind new friend turned out to be a volunteer warden, Roger Smith, who monitors the progress of the reintroduced butterflies by counting their eggs. Though only the size of pinheads, and well hidden on sprigs of thyme, the eggs are actually easier to find than the butterfly. Roger agreed that this butterfly can be remarkably fickle. It had settled into some of its introduction sites but not others and the reason wasn't always clear. All contained the black ant, *Myrmica sabuleti*, in whose nests the Large Blue caterpillar completes its development, and all contained plenty of wild thyme in a suitably short turf facing the sun. The good news was that the butterfly was proving capable of looking after itself. One plucky individual had flown over a sizeable wood to lay its eggs on the grassy hilltop beyond.

At last we spotted one: a male. Just one. It was jinking close to the ground and for a moment it decided to pick a fight with a much larger Meadow Brown. In those few seconds I noticed how dark the Large Blue appears on the wing – the colour of blue-black ink – dark enough to explain its older name of Dark Blue. It was also smaller than I'd expected, no bigger than a Chalkhill Blue. In an instant, the butterfly flicked its wings, caught the sun in a stab of light, and was gone. 'I'm not sure whether I can count that,' I remarked, rather ungraciously.

By now several of us were combing the hillside, peering closely at the ground for those sudden flicks of inky blue. Our second Large Blue was a female, more sluggish than the darting male, dove-grey beneath with larger spots and a delicate flush of silvery blue near the body. Suddenly and thrillingly, it settled on a head of thyme, and curled its concertina body under the buds, probing here and there. But evidently the plant was not to its liking for soon it stopped, tensed, and, in an inkling – a word that might have been coined for the Large Blue – it was gone. 'Did it lay?' The egg is a pricked porcelain dot with a hint of blue. I looked: 'No.' Quite apart from the temperature and the time of day, the Large Blue is fussy about egg-laying too. I was starting to get the impression that these ex-Swedish butterflies don't actually like England very much.

As I was leaving, thrilled at having seen the thing at last but a little disappointed to have spotted only two, I saw a plump, sweating man shouldering a camera on a tripod. He had come all the way from Lancashire. Our eyes met. 'I'll see it, won't ah?' he gasped. I told him that I'd spent the last three hours looking for it and, with a lot of help, had seen precisely two. But, I added brightly, it might get a bit livelier when the temperature drops. 'Oh . . . ah.' I could see he had hoped for better. I hope he got his photograph. It was a long way home to Rochdale.

If the Large Blue is British again, how 'British' is that? Our extinct form of the species was described as a distinct race, *Maculinea arion eutypon*, and, if you believe that, then our distinctive form of it has gone for good. It can never be reintroduced. Specimens in collections also suggest that our Large Blues varied more than most butterflies. Those from the Cotswolds were consistently dark and, in the case of the female, sometimes suffused with dark brown scales like those in the Swiss Alps. Large Blues from Northamptonshire were also dark whilst those from Cornwall and North Devon were paler, closer to the 'iron-blue' colour of the Swedish form. Evidently genetic isolation had produced a number of distinct forms of the butterfly each closely adapted to its particular area. This genetic adaptation to different local conditions, which must have evolved over centuries, is no longer valid. The more uniform butterflies from Sweden will have to find their own way, if they can.

The reintroduction of the Large Blue is often described as one of the great conservation success stories of our time. And so it is, up to a point, and those who made it possible deserve all credit for their determination and skill. Whether that success will endure without continuing care and attention is more questionable. The impression I took away from Collard Hill was that of an unusually pernickety butterfly, one that seems to fly only when things are just right: the right temperature, the right time of day, possibly the right mood. In the heart of their range, in central Europe, the Large Blue is nothing like so shy (it is bigger too). This is probably the most closely studied and monitored butterfly in Britain – and the most expensive. Everything that can be done for it is being done. But you still feel that this is a foreign butterfly not yet fully acclimatised to British conditions. Meanwhile dedicated bands of volunteers will go on planting thyme and counting eggs and doing their best to get what conservationists call 'the grazing regime' right, in the hope that one day the Large Blue will become better established than it is at present. I think we had better cross our fingers.

THE FINAL CURTAIN

Extinction is nature's last word. It is all about finality. All life forms on the planet will one day become extinct. In the life book of a species there is no surprise ending: it is always the same. Every one of them is in the position of the poor player in Macbeth who 'struts and frets his hour upon the stage and then is heard no more'. Extinction happens when the environment can no longer support the species. Forms better adapted to changed circumstances replace it; life goes on, and, on the whole, biodiversity is maintained. Just occasionally, a mass extinction wipes out a large proportion of life on earth. As everyone knows, the last really big one happened 65 million years ago when a very large lump of space-rock hit the earth and made an almighty bang. That was the end of the dinosaurs and pterodactyls and ammonites and much else besides. We are probably living through another of those catastrophic events, caused this time not by an asteroid but by human activity. We are already the greatest destroyer of life the planet has ever seen, and unfortunately we have barely got started. Wiping out species seems to be the destiny of mankind, whether or not we end it all by wiping ourselves out: the executioner executed.

All the same, we do not glorify ourselves as destroyers of life. In fact, extinction makes us feel bad. It smacks of failure. Fortunately our memories are short and only the best-educated British entomologist will remember such lost delights as the Orange-spotted Emerald dragonfly, wiped out when we polluted its river, or the Reed Tussock moth, the victim of a nationwide desire to grow carrots on our last wee wilderness. It is just as well that most British insects are resilient. Unlike species of primary habitats, such as rainforests, ours are man-adapted. They have found ways of surviving and even thriving in a landscape that is managed mainly to produce things we humans need such as wood and timber, milk and beef, pasture and hay. Most insects seem to have found a niche for themselves in this remade world, half-natural, half-artificial. The exceptions are those whose habitat has vanished altogether, such as those that were confined to the undrained Fens, or, like the Large Blue, depended on a finely tuned set of circumstances.

Butterflies may be more extinction prone than most other insects, especially in cool, wet, changeable places like the British Isles.[3] National extinction is our ultimate ecological tragedy. We have lost at least 500 species during the past couple of centuries. Yet positive things can happen even from the most negative event. The extinction of butterflies helped to give birth to one of the twentieth century's great ideas: conservation.

THE FIRE BUTTERFLY

The first British butterfly to die out was the Large Copper. Our Large Copper was of a large and distinctive race called *Lycaena dispar* subspecies *dispar* (the name *dispar*, meaning 'different', refers to the contrasting sexes of the butterfly: the bright and shiny male, the darker and spotted female). By a quirk of evolution, isolation among the fens of Lincolnshire and East Anglia had produced a jewel of a butterfly. Its country name was the 'fire butterfly', the one whose wings caught the light like flame, like a spinning penny.

Its beauty, as well as its rarity, made this butterfly intensely desirable. The coming of the railway network in the 1820s enabled London collectors to take a day ticket to Holme or Ramsey, in Cambridgeshire, within striking distance of the butterfly's marshy heartland. A lively trade sprang up between the fen folk, always happy to earn a few extra shillings, and the (as they supposed) wealthy collectors from down south. The plump green caterpillars were picked like blackberries from the tall water docks growing in the dykes, and sold in matchboxes for sixpence each. But as usual it was the middlemen who made the money. One of them bought two dozen young larvae off an old woman for ninepence, bred them up and sold the emergent butterflies for a shilling each. The butterfly itself was, it seems, hard to catch. Though bold, and willing to 'attack any insect that came near', it was also adept at dodging the net and, if you missed it, the fire butterfly seldom gave you a second chance. It was slippery work too, navigating your way over the dykes, pole in one hand, net in the other, or pursuing the butterflies through tall, dewy grass and sedges that cut like knives.[4]

Blaming the collector, as many did, simply ignored the obvious. While the marshes remained, there were always plenty of Coppers to collect: this was a sustainable harvest. But the moment the land was drained, the butterfly vanished, along with all the other insects that depended on the wet and the wild, whether they were collected or not. Today, most of the places where the Large Coppers had flown are now dry crop fields through which narrow lanes zigzag their way to lonely farmhouses on the horizon. Long ago, the wetness was drawn from the fens by steam pump, and consigned to an intricate web of dykes and lodes which flushed away the unwanted water into the Wash. The fire butterfly didn't have a chance.

Surprisingly, perhaps, butterfly lovers took the loss on the chin. The standard butterfly book of the day was by the Reverend Francis Orpen Morris, a conservationist *avant la lettre*. He campaigned against fox hunting and vivisection, and is said to have invented the bird table. But when it came to the Large Copper he found it impossible to gainsay the advantages of super-drainage. Its loss was a pity, of course, but it couldn't be helped. 'Science,' he wrote in one of his many pulpit sermons dressed up as natural history, 'with one of her many triumphs, has here truly achieved a mighty and valuable victory, and the land that was once productive of fever and ague now scarcely yields to any in lowland England in the weight of its golden harvest.' Food, in other words, must always come before butter-flies. No loyal and patriotic subject should 'repine in the face of such vast and magnificent advantages', continued Morris. And if he could take its loss with Christian fortitude, so should everyone else.

The handsome Yellow-crested Spangle butterfly, *Papilio elephenor*, is a large, dark swallowtail with flashes of pink and blue (but not, funnily enough, of yellow) found in north-east India. Once thought to be globally extinct, it was found alive and well in its old haunts after the passage of nearly a century.[5] If a butterfly as large and bold as a swallowtail can hide from mankind for a hundred years, how easy must it be for the hundreds of much smaller butterflies that live in the great forests of the world?

We have no idea how many butterflies have died out since mankind began to develop the wild places of the earth. It might be hundreds, or even thousands. But if so they disap-peared before anyone could document them. One of the few well-attested cases was the Xerces Blue, *Glaucopsyche xerces*. It was North America's equivalent of the Large Blue, and, like our butterfly, it acquired posthumous fame. Named after the Persian king who fought the Spartans at Thermopylae ('Xerces' is the French spelling of the more familiar 'Xerxes'), it was a small, darting butterfly whose colour matched the clear Californian sky. Nabokov described a similar species as 'celestially innocent'. Its caterpillars fed only on the native silver lupin of San Francisco Bay and so it was bad luck that the best colonies of the flower happened to grow on land scheduled for urban development. By 1943 there were no more natural dunes, no more lupins and no more Xerces Blues. Some believe that the *coup de grâce* was an invasion by a new species of ant that replaced the one on which the butterfly depended. All that is left of the Xerces Blue lies inside three drawers of pinned butterflies in the collection of the Academy of Sciences at Golden Gate Park. No one will ever again enjoy those living spurts of lapis lazuli that once darted above the white Pacific sand and settled with spread wings on the matching blue spikes of lupin.

Yet the name of the Xerces Blue lives on in the Xerces Society. Formed in America in 1971, it campaigns for insects worldwide, in its crisp, encouraging words, 'by harnessing the knowledge of scientists and the enthusiasm of citizens'.[6] Like the Large Blue, the Xerces Blue has become a talisman for insects in danger. A sad story of loss has thereby been transformed into one of hope: hope for another rare butterfly, the Coastal or Bramble Green Hairstreak, *Callophrys dumetorum*. Similar to our own Green Hairstreak, with brown topsides and pretty green underwings, it is confined to the remaining and now isolated patches of wild scrubland in the San Francisco area. Understanding now that habitat fragmentation spells doom for a species, the city's citizen army has been busy planting roadsides and 'street parks' with the butterfly's food plants, buckwheat and deergrass. San Francisco's 'butterfly corridors' have become famous as an example of what can be done to preserve a butterfly even on the streets of a large conurbation.[7] Providing a wild niche in the heart of the urban jungle – Virgil's *rus in urbe* – is, so the Xerces

Society argues, part of San Francisco's greatness. Aware of the past, it is willing to try and make amends. Civilisation must include conservation, for without it we are mere destroyers of the earth.

EVOLUTIONARY IRONY AT WORK IN MADEIRA

Maybe the Xerces Blue was just unlucky. It was a native Californian with a small home range, isolated behind mountain walls. Most butterflies have much larger ranges and so a far greater chance of finding a refuge somewhere. The most vulnerable species are those which have evolved on isolated islands and have nowhere else to go. This is what happened to the only known European butterfly to have vanished off the face of the earth. Ironically it was a close relative of the Large ('Cabbage') White, possibly the commonest butterfly in the world. It was confined to the island of Madeira and so was known as the Madeira Large White, *Pieris wollastoni*. Unlike the Large White, this one fed not on cabbages but on wild cresses that grow in the natural laurel groves of the island. Quite why it died out – the Madeira Large White was last reliably reported in 1977 – is uncertain. A possible cause was a virus infection brought to the island by the introduced Small White butterfly. Quite possibly the Madeira Large White was not a fully evolved species, for it differed from the Large White only in small details, notably the suffusion of pale yellow on the upper hindwings. Still, whatever it was, it was interesting and now it has gone.

The next one to go is likely to be Zullich's Blue, *Plebejus zullichi*, which is confined to the Sierra Nevada, an isolated mountain range in southern Spain, where a small population struggles to survive on a patch of hilltop.[8] One more nudge, from a new ski tow or car park, or a slight increase in overall temperature, and over it will go: one more tiny statistic on a growing list. Just behind it is a long queue of butterflies in danger. The survival of nearly 10 per cent of Europe's 482 species of butterfly is in doubt. A full third of European species are in decline.

POOR OLD SUFFOLK

The smaller the area it inhabits, the more likely it is that a species will die out. The largely urban island of Singapore has lost at least a quarter of its resident butterflies. By contrast, the much larger, only semi-urbanised island of Britain has lost only 10 per cent. That might sound pretty tolerable until you look more closely at the scale of a county. There, proportionate losses have been much greater. Suffolk, for example, has lost a full third of its butterflies. The pastoral landscapes painted by John Constable supported the Swallowtail, Large Tortoiseshell, Purple Emperor, most of the blues, all but one of the hairstreaks and all but two of the fritillaries. Most of them have since gone with, on the whole, little chance of coming back. Even the modest little Grizzled Skipper – a butterfly which asks very little from its environment – is in trouble.

The reason is the way intensive agriculture has changed the landscape of Suffolk from one of mixed farms with plenty of hedges, woods and flowery meadows to something closer to a prairie – a monoculture of wheat and barley. In the process, marshes have been drained, meadows reseeded or ploughed, and the old woods torn down or replaced by faster-growing, alien conifers. Other good spots were destroyed by roads and housing development. Suffolk's best-known entomologist, Claude Morley, saw it coming. Even back in the 1920s he was in 'no doubt that butterflies are dying out in this county . . . I do not mean that there will ever be no butterflies here but the ones you see will be merely the common kinds – the whites and browns and blues, the plebs of the highways and hedges' (just to be clear, the Latin word *plebs* means 'people'; plebeian means 'of the people').[9] By delicious coincidence the one scarce species to survive in Suffolk is indeed a 'pleb'. It is *Plebejus argus* ('small pleb with many eyes'), better known as the Silver-studded Blue. It flies defiantly in the shadow of Sizewell nuclear power station, representing, you might say, a victory of the plebs over all those butterfly aristocrats and emperors that have fared far less well.

BUTTERFLY JEWELS, LIVING DIAGRAMS AND THE TREE-DWELLING TORTOISESHELL

Mystery surrounds three more doomed butterflies: the lost British Mazarine Blue and the lost Black-veined White, and the is-it-lost-or-isn't-it Large Tortoiseshell. No one predicted their demise and no one could understand it when it happened. It was as if the names of all our butterflies had been shaken in a hat marked 'Extinct' and these three pulled out at random. If there is a lesson here it is that extinction usually comes as a surprise. Nature is full of them. And when the end comes it can be swift and can strike from a cloudless sky.

Today perhaps only jewellers know what a 'mazarine' is. It is or was a deep blue gemstone, a kind of sapphire, often worn round the necks of wealthy women in past centuries. The

Mazarine Blue is a living jewel, one of our darker blues with a sombre grey underside. Two hundred years ago it could be found, thinly spread, across southern England and Wales. Museum specimens of genuinely English Mazarine Blues are not common and so it might have been hard to find even then. It was a butterfly of damp clover meadows. One such meadow was a de facto nature reserve. Collectors paid the farmer an inducement to preserve the land for the benefit of the butterfly – and of course those who wished to catch it. But despite that, the Mazarine Blue died out there and everywhere else. The last recorded specimen was caught not by a collector but a spider. A Colonel Kershaw was walking near his home at Llanbedrog in North Wales in September 1905, when he spotted the bedraggled butterfly struggling in a web in the midst of a gorse bush. He freed it, placed the expiring butterfly in a matchbox, and took it home. It still exists, with a pinned label that tells the whole sad story. (Spiders *can* help to propel a butterfly to extinction: they were blamed for the demise of the Swallowtail at Wicken Fen in the 1950s.)

Thomas Chapman, who witnessed the disappearance of the Mazarine Blue, thought it was due to changing farm technology. So long as the hay was cut manually, with a scythe, there were enough refuges for the butterfly to raise a brood. But the replacement of scythes by horse-drawn mowing machines removed all the clover, eggs and caterpillars in one go. Jeremy Thomas has suggested that the butterfly might also have depended on ants to guard its caterpillars.[10] Relying on another insect is a high-risk strategy, as the Large Blue has demonstrated. Perhaps the wet summers of late Victorian England also helped tip the balance. We shall probably never know.

The Black-veined White died out a little later, in the early twentieth century. This was a much better-known butterfly than the Mazarine Blue. All the early books illustrated it. The artist Eleazar Albin had evidently bred it, for his engraved plate shows not only the butterfly and its caterpillar but also one of its parasites, a tiny fly. The Black-veined White is unlike any other European butterfly; its dark wing-veins stand out against a plain white background like a diagram, indeed, in the case of the translucent female, like a diagram on tracing paper. You could imagine the ancestor of all butterflies looking something like this. Today you have to visit the European mainland to see what English butterfly watchers once saw: crisp white butterflies thronging on wet seepages to imbibe the salts or at a communal roost on a flower stalk, like a cluster of large translucent seeds. It is a very sociable butterfly; Nabokov compared the sight of Black-veined Whites feeding in a puddle to 'little paper cockerels or a regatta of sailboats heeling this way and that'.[11]

In the nineteenth century the Black-veined White was found in southern England from Kent to the New Forest and in the west from the Cotswolds to South Wales. In some years it was common, in others scarce. Sometimes they were so numerous that collectors amused themselves by seeing how many they could scoop up with a single sweep of the net. The butterflies were particularly fond of moon daisies (or ox-eye daisies). It was a species everyone took for granted until, apparently quite suddenly, it wasn't there any more. In desperation, people tried releasing reared butterflies (the Black-veined White is very easy to rear). Whether a modest recovery at the turn of the century was due to that, or whether the failing population had one more try, it is impossible to say. Either way, the recovery did not last and the butterfly was effectively extinct in Britain by the early 1920s. The odd one has been seen since then, but, ominously, never any caterpillars – and this is a butterfly whose caterpillars were easy to find. Probably most or all records after 1923 were from releases, such as the industrial quantities of Black-veined Whites introduced into Winston Churchill's garden at Chartwell by L. Hugh Newman, all to no purpose as they did not survive.[12] This time no one blamed collectors. Some blamed the pheasant, others the wet summers of the 1870s and '80s. The full-grown caterpillars, it was said, caught cold and died in those wet Septembers. Again, its disappearance will probably always be a mystery.

So we lost the Black-veined White. And then we lost the Large Tortoiseshell. Many are reluctant to admit this – extinction is always hard to acknowledge – and there remains a small hope that this butterfly, which was always rather elusive, might still occur somewhere, perhaps on the Isle of Wight where there has been a tantalising series of reported sightings. This, the larger and rarer of our two tortoiseshells, was another well-known species: a handsome tawny butterfly as large and broad-winged as a Peacock with spots and bars of black, yellow and blue. It visited gardens and sometimes hibernated in the shed or in the crannies of a woodpile. Its powerful, soaring flight made the butterfly a hard one to catch. One mistimed sweep would send it up into the treetops. Some collectors bred it on branches of elm; at one time the Large Tortoiseshell was also known as the elm butterfly. Of course the death of nearly all of our tall English elms would have done it no favours.

The Large Tortoiseshell survived the dull weather of the late nineteenth century and thrived in the warm, balmy summers of Edwardian England. This is worth noting because the standard field guide for much of the twentieth century, South's *Butterflies of the British Isles*, was written at that time. According to South, the butterfly was 'more or less common in all the counties around London', especially in 'lanes margined with trees or the verges of woods'. There was another short-lived period of relative abundance during the hot summers of the late 1940s, but by then its range had contracted and most sightings were from East Anglia. It is impossible to say exactly when the Large Tortoiseshell joined the list of bygone butterflies but 1949 seems to mark the last time there was ever a reasonable chance of seeing one.

Once again there are many theories but no certainty. Perhaps, as a predominantly woodland butterfly, the Large Tortoiseshell fell victim to the increased shading of our woods in the post-war period, or perhaps it was afflicted by a parasitic fly in the same way as the Small Tortoiseshell was in recent years. Or possibly there was no single cataclysm but a slow accumulation of unfavourable circumstances that forced a dwindling in numbers until the population became unsustainable. We don't know because no one spotted the problem in time and no one investigated what was happening.

Could it happen again? Not if we can help it. When the Large Tortoiseshell died out, the study of insect ecology – that is, the relation of an insect to its environment – was in its infancy. Not a single species of British butterfly had been studied ecologically. Today butterfly ecology is a well-established discipline, worldwide, and the dynamics and requirements of butterflies are very much better understood. Perhaps today we would have known what to do about shrinking populations of the

Mazarine Blue and Large Tortoiseshell, and scraped enough resources together to preserve a large enough chunk of fenland to save the Large Copper. But the fear, if not yet the actuality, of extinction has not gone away. Not long ago the clever money was on either the High Brown Fritillary or the Duke of Burgundy as the next species to go. Ten years before that it was a toss-up between the Adonis Blue and the Silver-spotted Skipper. Today it might be the Glanville Fritillary. Tomorrow, who knows? Any one of a dozen species could die out. The challenge is to come up with ways of avoiding it. But there are many competing claims on our remaining wild land and some butterfly species are adapted to landscapes that are no longer commercially viable. For them it is conservation or nothing. The next chapter is about learning how to save butterflies.

12.

The Wall or How to Protect a Butterfly

The time when naturalists could write of 'clouds' of blues or 'snowstorms' of moths and expect to be believed are long past. Modern technology-driven farming leaves little room for butterfly habitats and our more exacting species have been driven to the margins, notably to the less-than-10 per cent of the land that has been given a measure of protection as Sites of Special Scientific Interest ('SSSI') or nature reserves. But compared with the past we have a much better idea of how many butterflies there are and how they relate to their environment. We have national distribution maps, repeated at roughly five-year intervals, and, for some places, more detailed maps too. We have information about their numbers from a network of 'butterfly transects' across the country where butterflies are counted along a fixed route by local volunteers. There have been detailed ecological studies of individual species. And there is now a successful charity devoted to preserving our butterflies and moths: Butterfly Conservation, based in Dorset, our richest butterfly county, with representative branches across Britain.

Butterflies are popular, probably more so than ever before.

Some 40,000 people took part in the annual Big Butterfly Count in 2013, a nationwide survey that sets out to 'feel the pulse of nature'.[1] Butterfly Conservation currently has 20,000 members – more than any other specialist wildlife charity except the RSPB. When I first became interested in butterflies, fifty years ago, it was all quite different. There was no butterfly society, no butterfly counts and only one or two butterfly books. Today there are not only dozens of books but websites, forums, blogs and films, and even butterfly holidays that take enthusiasts to some of the best butterfly places across Europe and beyond. Butterflies are no longer just a hobby but, for some people, a business.

Of course a fundamental task of any business is ensuring there is enough stock. Interest may be increasing but butterflies, unfortunately, are not. They are, in the words of Butterfly Conservation's latest report, 'one of the most threatened wildlife groups in spite of increased conservation expenditure'.[2] Just over half (54 per cent) of British butterfly species are going downhill – a proportion that increases to two-thirds if you include falling numbers as well as shrinking range. (On the other hand one-third (31 per cent) of our butterflies have managed to defy augury and are actually commoner today than twenty years ago.)[3] If butterflies are disappearing over the world at the same rate as in Britain – and, given the pressures from burgeoning human populations, they probably are – then we are facing entomological meltdown.[4]

In an ecologically perfect world, which I would define as a world with no people in it, butterflies would manage well enough, as they presumably did for millions of years. But our activities have twisted and pulled the natural world so out of shape that hardly anything is fully natural any more. Our more wide-ranging butterflies have adapted to man-made habitats such as gardens or railway sidings or brownfield sites noted for their ragged jungles of naturalised buddleia and other nectar-rich flowers. But others are still more or less tied to patches of natural habitat such as uncultivated down and heath, coppiced woodland and natural coastlines. Our smallest butterfly, the Small Blue, seldom strays more than a few hundred metres from the patch of kidney vetch where it was born. When the Small Blue 'returned', as *Butterfly* magazine expressed it, to the Ayrshire coast, having died out there thirty years ago, it was from butterflies introduced from a donor site more than a hundred miles away.[5] There was apparently no suitable donor site nearer than that.

We like butterflies and we want to make sure they survive. How do we set about it? Nature reserves present one answer: they supply wild plants and animals with a supposed sanctuary where they are at least safe from tractors and bulldozers. Unfortunately nature reserves do not have a brilliant record of preserving butterflies. Many rare species have died out in their supposed sanctuaries – to the great frustration of those who set them up. What was missing was the kind of management that created and maintained small-scale butterfly habitats. The Adonis Blue cannot exist without sheep or rabbits to crop the turf and warm the soil. The Heath Fritillary cannot survive under the cool shade of packed trees. The Marsh Fritillary seems to demand the kind of finely tuned grazing that has you reaching for your pocket calculator.

Conservation has given a new cultural meaning to butterflies. Past generations admired them, wondered about their lives, painted their pictures, pinned them in collections. Today, while still admiring their beauty, we tend to see them more as symbols of fragility, eco-victims in the casualty wards we call nature reserves, perhaps. Conserving butterflies in grossly overpopulated, intensively farmed Britain presents a challenge. The first thing it needs is knowledge. That box, at least, has been fairly thoroughly ticked; probably more resources have gone into the study of butterflies, especially British butterflies, than any other group of insects on earth (the purpose of all this study is, of course, to conserve them, the opposite of traditional commercial science which is to get rid of pests such as greenflies or mosquitoes). We can probably tick off determination too. Lots of people care a lot about conserving butterflies. There are also resources available, not as much as there used to be when the economy was more buoyant, but sufficient to make some progress towards the aim of providing genuinely safe refuges for our most endangered species.

But butterfly habitats are still being lost at an unsustainable rate. The natural landscape is too fragmented to provide large, safe refuges, and managing smaller sites for butterflies is often problematical. It may take an awful lot of planning and physical effort just to maintain the status quo. You could compare it to one of those running machines in a gym: you may be running, you may be out of breath, but you are not moving forward. Looking after the wild landscape is like that. Grassland wants to turn into scrub and then woodland. Nature conservationists might want it to stay as natural grassland (which is a lot rarer than natural woodland). To achieve that requires farmers, sheep, cows, mowers, barns, fences, volunteers, planning, and, above all, money.

*   *   *

Until the 1970s, our knowledge of British butterflies had not advanced much since the reign of Queen Victoria. Successive books rehearsed the same old facts of distribution, life cycle, food plants and 'aberrations' but very little was known about how butterflies survive in the wild, about how their numbers are regulated and how they interact with other species and with their surroundings. We didn't know much because hardly anyone was looking at living butterflies. That is perhaps the fairest accusation you could level at the collectors: they were too busy pursuing specimens to watch patiently and try to understand what the butterfly was doing.

We have come a long way since then. Martin Warren, head of Butterfly Conservation, sees the story as a succession of breakthroughs, beginning with the publication of *Silent Spring* by Rachel Carson in 1963.[6] In response to public anxiety about the side effects of pesticides such as DDT, a research station was set up to study what these poisons were doing to wildlife. This was the famous Monks Wood experimental station, near Huntingdon, named after the wood next door – which, as it happened, was one of the few nature reserves set up specifically to protect butterflies. And although butterflies were not at first explicitly part of Monks Wood's agenda, it was natural enough that some of its ecologists began to take an interest in them.

One of them was John Heath, a keen field entomologist who was in charge of biological records. He organised the first systematic survey of butterflies that resulted, in 1984, in the first butterfly atlas.[7] This revealed for the first time how badly many butterflies were faring. I still remember the shock of seeing all those open circles on the map, marking the places where butterflies once flourished but were now gone. You only needed to go to Monks Wood for a sharp example. Of the forty-one species of butterfly that once flew there, eleven had vanished, including many of the species the reserve had been set up to protect. If this could happen to a National Nature Reserve, the highest form of protection in the land, what chance did butterflies have on ordinary farmland or in the commercial forests?

Another idea pioneered at Monks Wood was the 'butterfly transect'. Ernie Pollard had studied Swallowtails and White Admirals, more or less as a hobby. It was he who worked out why Swallowtails no longer flew at Wicken Fen: the marshland had partially dried out and the milk parsley, on which its caterpillars fed, grew less luxuriantly. The result was increased predation and fewer butterflies until the population became unsustainable. Simple ideas are often the best. Butterfly transects were simple. Botanists are used to monitoring wild flowers by means of a fixed line called a transect. Pollard reasoned that butterflies could be counted and recorded in the same way by extending the transect into a fixed route along a path so that butterflies could be counted during a short walk of an hour or two. By repeating the exercise at regular intervals you could acquire a simple set of data that could be compared with other transects, and from one year to the next. Anyone able to identify butterflies could take part and the scheme was tailor-made for amateur naturalists who were no longer collecting butterflies but eager to take part in an organised recording scheme. In time Pollard's idea ripened into the UK Butterfly Monitoring Scheme. A measure of its success is the quarter-million weekly visits to 1,500 monitoring sites across Britain and Northern Ireland totalling half a million kilometres. Since the scheme began in 1976,[8] more than 16 million butterflies have been recorded on butterfly transects.

The third breakthrough might be called the Coming of Jeremy Thomas. He was the first to study, for a PhD thesis, the ecology of British butterflies in the field and in doing so virtually invented this now flourishing science. Thomas's chosen subjects were the Black and Brown Hairstreaks, rather mysterious butterflies which are notoriously hard to observe and about which little was known. They share a common food plant, blackthorn, but Thomas was able to show that their life strategies are quite different. The Brown Hairstreak is the more widespread of the pair, occurring at low density across much of southern England and Wales where it lays in blackthorn hedges and thickets in the vicinity of woods. The Black Hairstreak, on the other hand, prefers old, overgrown blackthorn along the edges of woods, notably on those skirts of scrub that you find in mature woods on clay soils. That discovery went some way to explaining the Black Hairstreak's curiously limited distribution, more or less confined to the belt of heavy Midlands clay between Oxford and Peterborough.[9] The butterfly is perfectly capable of living elsewhere as was proven when it was introduced into a wood in Surrey and thrived there. But it is not likely to find such woods on its own because this is one of our least mobile species, rarely straying far from the small wood of its birth. Thanks to Jeremy Thomas's work, we know what to do to help the Black Hairstreak and for once it is pretty simple: preserve its habitat. Because it is relatively easy to preserve woodland scrub, the Black Hairstreak, of all our rarer species, is the one that has declined the least. It even survives at Monks Wood.

Inspired by Jeremy Thomas, others chose butterflies for

their doctoral studies. Martin Warren's own thesis was on the Wood White, another threatened woodland butterfly, and he followed it up by an equally intensive study of the Heath Fritillary. This too led to more effective conservation. I still remember my first encounter with this small, bright fritillary in the Blean Woods near Canterbury. I thought I was dreaming. What looked like hundreds of them were gliding, fluttering and settling in the sunlit glade above a woodland floor crowded with the yellow heads of the butterfly's larval food plant, cow-wheat. Conservation had come to the Heath Fritillary in the nick of time. This is a woodland butterfly but paradoxically one that demands plenty of sunshine; it used to 'follow the woodman' as they cut the chestnut coppice to make paling fences. It was the woodcutters who created the open conditions this butterfly needs and their absence that threatened it when Kent's coppice industry hit the buffers. Conservation volunteers have taken the place of the woodmen and, under the swing of their billhooks, the Heath Fritillary is thriving again in a few woods in Kent and Essex, as well as in more open country in south-west England.

Such studies provide the intellectual bedrock around which conservation decisions can be taken. Keith Porter's meticulous work on the Marsh Fritillary, for example, revealed that it is the needs of the caterpillar more than the butterfly that demand attention (and recalling the remark of the American comedian George Catlin that 'the caterpillar does all the work but the butterfly gets all the publicity').[10] The caterpillar is the most vulnerable stage of a butterfly's life cycle. It must grow as fast as possible to escape its many predators. In human terms they need to be fit – full of energy and vitality to achieve their single goal of feeding up as fast as possible. To do this the caterpillar must stay warm and it does so by alternating its feeding frenzies with basking in the sun. When I used to rear butterflies I experienced high losses with some of the fritillaries, and the most likely reason is that my cages were not warm enough; my poor, cold caterpillars were denied their regular sunbathing. It is the same in the wild: caterpillars are not going to get warm in dense, damp vegetation or dewy long grass. They need dry, dead leaves in full sun. Such conditions appear when the ground is lightly grazed, in the case of the Marsh Fritillary preferably by hardy breeds of pony or cattle. In the remote past perhaps the job was done by wild aurochs and other big, now lost wild animals. Porter's work showed that you need to be conscious of the needs of the early stages quite as much as, or even more than, the adult butterfly.

A fifth breakthrough was the discovery that butterflies are very fussy about where they lay. It is not simply a matter of finding the right plant. The turf must be the right height and the vegetation must be sufficiently open and sunlit. To the butterfly it must 'smell' right. You watch a female butterfly inspecting a leaf or bud carefully, begin to curl its abdomen to lay, and then seemingly change its mind and fly off to find some better plant. It seems to be able to detect the presence of chemicals such as amino acids that are necessary for the growth and development of a healthy caterpillar. Purple Emperors, for example, seem to prefer fresh green sallow leaves to older, perhaps more bitter-smelling ones. Temperature is also important. Grass-feeding species such as the Wall and the Silver-spotted Skipper seek not just any old grass but short open turf with the soil showing through; collectors used to say of the skipper that it was no use searching for it until you could see pebbles of chalk among the grass.

Martin Warren's last, most recent breakthrough is also the most technical. It involves the concept of 'metapopulations'; that is, separate populations of butterflies which interact to some extent and so form part of a much larger unit – not unlike the separately named villages that make up the greater conurbation of Birmingham. The individual 'villages' are expendable; it is the larger metapopulation, the butterfly 'city' that must be conserved at all costs. The idea was worked out most thoroughly by the Finnish ecologist Ilkka Hanski who used the Glanville Fritillary as his object of study in Finland's Åland Islands. From his long run of data, Hanski proved that individual populations come and go in an agricultural landscape, but so long as there is some interaction, some migration, between one colony and another, the greater population has the capacity to heal itself. It can in time recolonise lost sites.[11] Hanski's work has been enormously influential to those who plan conservation strategies. Not only has Butterfly Conservation nailed it to their mast but the idea now lies at the heart of government policy on saving biodiversity.

Personally, I would add one more breakthrough: Martin Warren himself. He was the Nature Conservancy Council's first butterfly specialist (the NCC was the forerunner of Natural England, the government's advisory body on wildlife and nature conservation). His knowledge and advocacy were hugely beneficial to the way nature reserves were managed for butterflies and other insects. From minor players butterflies became lead actors on the conservation stage, especially on chalk grassland and heath. In 1993, Martin joined the then small charity,

Butterfly Conservation, as its first conservation director and is now its chief executive. He is also an authority on European butterflies and co-author of the *Climate Risk Atlas*, a detailed assessment of the likely consequences of a warming climate on butterflies. In 2007 he won the Royal Entomological Society's Marsh Award for Insect Conservation and was voted one of the top ten most influential conservationists in Britain. Under his leadership Butterfly Conservation has grown from a minor to a major wildlife charity with an annual turnover of £3 million, as well as a fine team of full-time conservationists, an army of volunteers (whose work is worth an estimated £10 million annually) and the best magazine in the business. Butterfly Conservation looks after the 'action plans' of upwards of sixty rare butterflies and moths, and hosts international conferences on butterfly science. In short, Martin is the best ambassador British butterflies have ever had. And I'm sure it is not entirely irrelevant that he is one of the nicest people under the sun.

The more we discover about them, the more exacting butterflies seem to become; in other words, the more you know, the harder it gets. The year 2013 was a good year for promoting butterflies but a bad one for funding. Conserving rare butterflies is expensive; it takes up two-thirds of the charity's budget and even that is less than a tithe of what is necessary simply to hold butterfly numbers at roughly their present rate. Butterfly Conservation depends on government (Defra) handouts for its more ambitious projects and finds its budget knocked sideways every time the government trims the department's spending – and lately this has become a routine annual scything.

Managing a piece of land in the interests of a rare species

can be a formidable undertaking. Take one nature reserve in North Devon. Though an interesting place for wildlife generally, its 'flagship' species is the Pearl-bordered Fritillary. That butterfly is the yardstick of success or failure. Saving it and its slightly commoner cousin, the Small Pearl-bordered Fritillary, is the number one priority. Butterfly Conservation shares the burden with Devon Wildlife Trust. They aim to restore a scrub-invaded valley to its former glory, a place of grassland and heath mixed with mature woodland. And that needs a lot of work. All the cutting, burning, raking and winching it involves would take several pages to describe. Instead ponder the implications of this one short paragraph from a lengthy article about the restoration work in progress:

> Much of the cut material [bracken and gorse] is raked and removed . . . although in some areas bracken is left as a light cover . . . Raked material is stacked into piles. Some of the cut scrub is burnt and about 10% of the stumps are treated. Additionally, some bracken sections are cut monthly during the summer season in order to create open areas for use by other butterflies . . . Bracken that encroaches onto grassland areas is also cut during the summer. The height of cutting for bracken and gorse compartments is about 10cm and is never allowed to grow level . . .[12]

And that is just the bracken. At any particular moment you might see a tractor cleverly adapted for rough country pulling a mounted swipe or a mower, or listen as the buzz of a saw or the whine of a winch drowns out the birdsong. You might spot three or four Exmoor ponies doing their unconscious bit for butterflies by browsing the gorse and grazing the tough grass. You might notice plumes of smoke drifting from the glades as the brushwood is consumed or a patch of turf deliberately scorched to encourage violets on which the woodland fritillaries depend. All this activity requires two full-time workers (the warden and his assistant) aided by voluntary work parties, summer and winter. Some of the funding comes from the Forestry Commission, some from the Landfill Communities Fund and the rest from the Devon Wildlife Trust. Progress of the colonies of fritillaries has been mapped by the square yard.

Projects like this one depend on a keen sense of commitment and enthusiasm from local volunteers. Each place where men and women are busy cutting and mowing and burning to sustain butterfly habitat represents its own small contribution to the future of butterflies in Britain. But you cannot help wondering whether this is really the long-term answer. It is wonderful that so much can be done but daunting that so much work seems to be necessary. A nature reserve might need as much or more management than a small country estate, and all for a butterfly. And it must be ongoing, without pause, year in, year out, if you are not to risk losing species to the inevitable return of bushes, brambles and coarse grass. This can only be a solution in certain places where there is enough money and enough volunteers to do the job.

Some might say that if the landscape has really turned so decisively away from the needs of certain butterflies then it might be more honest to let them die out, if that is to be their fate. They belong, you might say, to the landscape of yesterday: of Thomas Hardy's heaths or John Constable's elm-fringed meadows. Maybe the next generation would think no more of the lost Pearl-bordered Fritillaries and Dukes of Burgundy than we do about half-forgotten Black-veined Whites and Mazarine Blues. The answer, of course, is that without sympathetic land management we would lose a significant proportion of our present biodiversity, and not just butterflies. But even conservation charities are starting to recognise that nature reserves – butterfly ghettoes – are almost a council of despair, a procedural blind alley offering short-term expediency rather than a long-term answer.

The wall which separates nature reserves from the rest of the countryside is wholly artificial. It is, some would say, yesterday's solution, and it is not working. Some say the wall should come down. Hope lies rather in the new idea of metapopulations. If enough of the countryside could be managed in a moderately butterfly-friendly way and perhaps linked up somehow, then butterflies and other wildlife might survive through their own resources instead of by expensive and ultimately futile action-planning. But that is a very big 'if'.

These are deep waters. What is actually achievable in this nervous new world of financial restraint and climate change is uncertain. At the dawn of the new Millennium, Butterfly Conservation proposed a ten-point plan to save our butterflies from further decline.[13] It wanted a reform of the EU Common Agricultural Policy to shift payments from production to environmentally friendly farming. It urged government to be more generous in funding conservation such as the Species Action Plans for rare butterflies and moths. It wanted stricter protection for species, better protection for important butterfly sites and better policies for promoting biodiversity. Five years on,

Butterfly Conservation gave marks for progress: six out of ten for agricultural reform, seven out of ten for sustainable development and five out of ten for government funding. Today, with a government committed to deep cuts, many would find those marks to err on the generous side. The 2010 EU target for butterfly conservation was manifestly not met – and so, in the way of things, the EU has set itself another target although there is no indication, so far, that things will look any rosier in 2020 (perhaps the purpose of targets is to represent aspiration – to offer a goal rather than an honest prediction). In the meantime conservationists are going to have to become less reliant on central funding. The old expensive ways of managing the environment no longer apply.

How would it look if we lost another butterfly or two – the two pearl-bordered fritillaries, say? How would we feel about that? After a lifetime of floundering against the tide in the conservation business, I have grown fatalistic about our ability to prop up wildlife against the remorseless flood of population, houses, roads, and sterile agriculture. I agree with the *Independent* journalist Mike McCarthy who, in his farewell piece after forty years in newspaper journalism, concluded that it is the destiny of mankind to destroy nature. But even Mike conceded that we can put off that fateful day by greener policies. Conservation thrives on hope just as a plant needs water, for without it there seems no point in even trying. Contemporary language enjoys aspiration; every ghastly circumstance is palliated as a 'challenge', every unlikelihood with the word 'hopefully'. But conservationists have also learned to be pragmatic; they are usually willing to accept, even to praise, a glass half-full, or even quarter-full; it is better than nothing. Even HS2, the High-speed Railway, is, for some, an opportunity for 'rewilding' along the gaping wound it will carve through the Midlands and Home Counties. And who would not be an optimist on a warm summer's day walking on the downs with countless heaven-coloured butterflies – Robert Frost's 'sky-flakes down in flurry on flurry'[14] – dancing over the scented turf?

Nature would probably get on without butterflies. Unlike bees, they are not vital pollinators.[15] Unlike moths they probably do not have any significant role as bird (or bat) food. Unlike earthworms, they are not energetic recyclers. They have proved useful as 'model organisms' for science, especially in evolutionary studies and genetics, but in their absence other organisms would do just as well. The planet does not 'need' butterflies any more than it needs whales or hawks or primroses – or humans for that matter. But do we humans need butterflies? Butterfly Conservation thinks we do: they 'conjure up images of sunshine, the warmth and colour of flowery meadows, and summer gardens teeming with life'. They remind us of 'the essence of nature, or as representing freedom, beauty and peace'. Perhaps, in that sense, they also help to make us slightly nicer people. They certainly help to 'sell' the idea of nature, as a green life-style, as a haven of peace, as something precious and worth hanging on to.

When push comes to shove, it is hard to make an economic case for butterflies. We do not need them; nor, in the natural order of things, do they need us. But we care about them all the same, because that is the way we are.

13.

Envoi

Aurora or the Daughter of Dawn

Our English names for butterflies are, or should be, a source of pride but some of the French ones are just as inventive: *Souci* (marigold) for the Clouded Yellow, for example, or *Robert-le-diable* (Robert-the-Devil) for the reddish, ragged-winged Comma. The French name for the Orange-tip butterfly is *Aurora*, the goddess of the dawn. They see in its brilliant orbs the rising sun, fresh and beautiful, glowing on the horizon. Perhaps it is surprising that the English missed this allusion, for those who named our butterflies were amateur poets well versed in mythology and classical allusions. We did once have a better name for it: the Wood Lady, or the Lady of the Woods. Unfortunately the butterfly with the sunrise in its wings is no lady; it is the male of the species (the much less conspicuous female has no orange tips to its wings) and it lives in wet meadows and waysides more than in woods. But at least the scientific name, *Anthocharis cardamines*, does justice to the Orange-tip. *Anthocharis* means flower-grace, and what could better describe this pretty, dancing butterfly as it patrols the hedgerow 'searching every shrub and tussock for a mate'.[1]

The Orange-tip represents the dawn in another sense. It is one of the first butterflies of the spring, and when you see one, then several more, then, with luck, dozens, you know that winter is over and that the blossoming and greening of spring is with us at last. It welcomes us to a new beginning, nature's wakening.

Summer 2013 was, in the end, a wonderful year for butterflies. It was slow in coming; spring was cold and wet; early summer disappointing. There were even fewer butterflies than the previous year when it seemed to rain non-stop from April to August. But, towards the end of June, the skies cleared and what the weathermen call the Azores High was at last on its way. July 2013 became the third-warmest month in the past hundred years, some 2.5 degrees C above the average. We enjoyed six weeks of sunshine. By autumn it had turned into the best summer for seven years; one of the best of the new century.

And the butterflies responded. Brimstones and Peacocks feasted on garden buddleia. The Small Tortoiseshell, whose numbers were so low over the previous decade, suddenly returned in force. Large and Small Whites also flew in numbers rarely seen in recent years. Someone on the North Norfolk coast witnessed coming in from the sea a 'cloud' of white butterflies whose numbers he estimated at 40,000. Someone else, walking a butterfly transect in Sussex, had to give up counting the 'shimmering' Chalkhill Blues for there were simply 'zillions' of them. The Purple Emperor was spotted in places where it had been absent for decades. And it was also the best year ever for that very rare migrant, the Long-tailed Blue. Usually recorded only in single figures, if at all, it was seen by many people along the south coast from Devon to Kent, and above all in gardens. Almost for the first time, a British-born generation of Long-tailed Blues appeared in September. This pretty butterfly, its wings a harmony of shimmers, bands and eye-spots, further distinguished by a pair of delicate, whiskery 'tails', may be on its way to joining the Red Admiral and Painted Lady as a regularly breeding migrant. It makes a statement: the weather is getting warmer. The south is moving north.

The butterfly bonanza of late summer 2013 lingered as late as September and it seemed to lift many spirits. You sensed it from the blogs and twitters of butterfly watchers like Martin Warren and Matthew Oates. Slightly despondent in May and June, with the coming of warm weather their messages turned into a kind of thanksgiving. 'Days like this do the soul good,' tweeted Matthew Oates. 'Heaven has decided to come down to earth,' and, quoting Ruskin, he evoked '"the poetry of Nature . . . which uplifts the spirit within us"'.[2]

Summer 2014 was just as uplifting but in a different way. Though actual numbers rarely matched the previous year, the unscheduled appearance of two spectacular species gave notice of further change. The immigrant, continental form of the Swallowtail arrived on the south coast in numbers unprecedented since the hot summer of 1945. Some of them laid eggs that hatched into gaily striped caterpillars which fed up on fennel in gardens and waste places. Some of them survived to turn into chrysalids and some of those chrysalids hatched into British-born Swallowtails the following spring. Like the Long-tailed Blue, the continental Swallowtail may be in the process of turning British. A Swallowtail in the garden would feel like a holiday. Perhaps we will all see them soon.

It was also the year of a completely unexpected species – the Scarce or Yellow-legged Tortoiseshell. This is another colourful, strong-winged butterfly, very similar to the lost Large Tortoiseshell but distinguished by its pale legs and a white spot near the apex of the forewing. Hitherto almost unknown in Western Europe, let alone isolated Britain, it arrived in small numbers along the coast of eastern England during July 2014. Single butterflies were also spotted in Devon, Tyneside and the West Midlands. The survivors among them will attempt to hibernate like the sleepy individual spotted in a cupboard under the family TV in Norfolk in September.[3] And, if all goes well, they will emerge again in the spring. The Scarce Tortoiseshell has always been a butterfly of comings and goings; in recent years it appeared in large numbers in Finland and Sweden, and lately it has also turned up further west, in the Netherlands. Whether we will see more stray Yellow-legs from across the North Sea, and whether some of them might lay and settle remains to be seen. There would be a poetic justice in it: having lost one large tortoiseshell, we would gain another, and solely by its own efforts.

Yet there is still hope for the Large Tortoiseshell. Its numbers seem to be building up in the Channel Islands and there have been springtime sightings of tattered tortoiseshells in the Isle of Wight, which suggest that they have been overwintering there. And if they spent the winter with us, there are presumably eggs in some as yet undiscovered place along the island's elm-rich hedgerows. This lost butterfly could yet return.

So, while we are losing butterflies, we are also gaining them. The stability that has marked British wildlife for centuries has gone, perhaps for ever. In the climatic turmoil that awaits us, some species, such as the Swallowtail, the Long-tailed Blue and perhaps the two larger tortoiseshells, may be able to take advantage. For those who live by the south and east coast, especially, there will be more excitements ahead to compensate, in some measure, for ongoing losses among our resident species. We will learn to live with new circumstances, and weave them into our sense of nature and the natural world. The inner butterfly that animates our feelings and sensitises our perceptions will live on regardless of the cancer-like spread of bricks, tarmac and concrete.

As I write, it is early autumn – a warm dry September of the sort that is still called an Indian summer, long after colonial memories of the Indian climate have faded away. It has fully lived up to Keats's 'season of mists and mellow fruitfulness' with the bees filling their honeycombs with the produce of 'more, later flowers'. When I walk up my lane on a still, warm evening the scent of ivy and mint hangs in the air. It draws you in, like expensive perfume; it is not, you suddenly realise, merely a scent but an attractant. Briefly you glimpse the butter-fly's world of invisible, chemical messages, in Miriam Rothschild's words, 'loading the night air and sometimes the sunny afternoon breezes with intense sexual stimulation and imperious desire'.[4] I pass the great masses of ivy tumbling over the wall in whose dark, windless recesses Brimstone butterflies will pass the winter. I wander into the village pub and sit in the bay window where a Victorian glassmaker has added some whimsical butterflies to his border of finches and sparrows. I watch the barman draw the cork of a new vintage with a bottle opener named after a butterfly – the Valezina, the dark variety of the female Silver-washed Fritillary – invented by a bygone collector, Edward Bagwell Purefoy, the man who first elucidated the mysteries of the Large Blue.

Edward Lorenz, the American mathematician, claimed that the fluttering of a butterfly's wing on one side of the world can create a hurricane on the other. He called it the butterfly effect. I have never understood it, and I don't really believe it. But I do believe that the same flutterings can summon up at least a small breeze in the human soul. Let it blow. Let it rustle the treetops and bend the grass and ripple the waters. Let the inward breeze remind we earthbound humans everywhere of the power and the wonder of the natural world.

British Butterflies

The usual way to present butterflies is by family, beginning with the primitive, moth-like skippers and ending with the 'browns'. Instead, and in keeping with the human-centred focus of this book, I prefer to begin with the commonest and most familiar species and finish with the rarest. These thumbnail sketches are not intended to be comprehensive. But within the small frame of each butterfly's personality I have included a few details of season, larval food plant and distribution.

Common Butterflies

LARGE WHITE, *Pieris brassicae*

Commonly known as the cabbage white, this is the butterfly that is most linked with human habitation. Gardeners and allotment holders will know all too well their clutches of yellow, bottle-shaped eggs and swarms of hungry, hairy, yellow-and-black caterpillars, busy perforating and shredding the leaves of brassicas and nasturtiums. The caterpillars imbibe mustard oils from their food plant which makes them distasteful to birds and mice. But this defence is of little use against parasites and many cabbage white caterpillars end up as a living dinner for parasitic flies and wasps. The Large White is a wanderer, easily capable of crossing the Channel, and can be found almost anywhere. Setting understandable antipathy aside, it is an attractive butterfly, its great white wings bordered and dotted with black and grey scales, and, beneath, pale yellow delicately speckled with black.

SMALL WHITE, *Pieris rapae*

This smaller relative of the Large White also feeds mainly on brassicas but its plain green caterpillar is more solitary, better camouflaged and hence not so familiar. It too is a very common butterfly and large numbers may build up by the end of summer. In the Great Butterfly Count of 2013 the Large and Small Whites were the most numerous of all our butterflies. Apart from the size difference, Small Whites are distinguished by smudgy dots on their forewings with less black on the wingtips. The smaller spring generation is a pure white; the summer one is overcast with black scales.

SMALL TORTOISESHELL, *Aglais urticae*

The lovely orange, yellow and black Small Tortoiseshell is not only our most colourful garden butterfly but also often the commonest. Among its favourite flowers are buddleia, ice-plant and Michaelmas daisy. The butterfly also enters houses looking for some dark, quiet place to aestivate (a kind of summer-sleep) or, later on, to hibernate. Emerging in the first warm days of spring, generally looking faded and tattered, the Small Tortoiseshell is one of the longest-lived butterflies. It is also one of four whose caterpillars feed on stinging nettles in sunny places.

PEACOCK, *Inachis io*

The Peacock, formerly known as the Peacock's Tail, has a unique set of iridescent 'eyes' set in each wing. Like the Small Tortoiseshell it is a frequent garden visitor, especially to buddleia flowers. Its spiky black caterpillars feed on nettles and are easy to spot when full grown. The Peacock is another long-lived butterfly, emerging in late July, hibernating in cool, dark places over winter and emerging again in the spring. It is one of the few butterflies that can make an audible noise – a faint rustle as it shakes its wings.

RED ADMIRAL, *Vanessa atalanta*

The Red Admiral is one of the world's best-known butter-flies. It is not a British resident but a long-distance migrant. Each year its splendid red-slashed wings power the butterfly from the Mediterranean to northern Europe where it lays its eggs on nettles to produce a fresh, British-born genera-tion. In recent years a few Red Admirals have survived the winter by hibernating. Hence the butterfly can be found at almost any time of year – even on mild, sunny days in January. It is commonest in late summer when the butterfly is attracted to rotting fruit in gardens and orchards as well as buddleia and Michaelmas daisies. Like all migrant butter-flies the Red Admiral is commoner in some years than others.

COMMA, *Polygonia c-album*

The Comma is easily recognised by the ragged outline of its wings and the tiny white 'comma' or c-mark on its dark under-side. No butterfly looks more like a withered leaf when at rest. The Comma is a butterfly success story; it is actually commoner and more widespread today than it was in Victorian times. There are two broods, a bright orange form in midsummer and a darker brood later in the year which overwinters. Like the previous three butterflies, the Comma lays its eggs on nettles, though the caterpillar also feeds on hop and elm. Unlike them, the Comma is seldom seen in large numbers but a patch of overripe blackberries in a sunny position is sure to attract a few.

BRIMSTONE, *Gonepteryx rhamni*

This bright yellow butterfly is a familiar harbinger of spring as it flies along the wayside looking for a mate. Only the more active males are sulphur-yellow; the slightly larger female is paler and in flight can be mistaken for a Large White. The Brimstone is a frequent visitor to gardens where it is drawn to pink or purple flowers. It lays only on buckthorn or its relative, alder buckthorn. The butterfly emerges from its chrysalis in late July and remains on the wing for a month or so before hibernating among ivy. It wakes up in early spring, usually still in good condition, and only then does it mate and lay its eggs.

ORANGE-TIP, *Anthocharis cardamines*

Another familiar springtime butterfly, the Orange-tip is a restless inhabitant of waysides and damp grassland. Only the male has the bright orange wingtips; the more elusive female resembles a Small White but has green scales on its undersides – colouring that makes a very effective camouflage when the butterfly is feeding or at rest. Its orange, bottle-shaped eggs are laid on hedgerow garlic mustard or lady's smock, both of which have long seed pods on which the slim green caterpillars feed.

GREEN-VEINED WHITE, *Pieris napi*

The Green-veined White is a much-maligned butterfly. Though related to the Large and Small Whites it rarely shows any interest in crop plants, despite its species name of *napi* – from *napus*, the turnip or swede. Like them, it is a common garden visitor but lays its eggs on wild cresses such as lady's smock and watercress in damp places and woodland rides. It is so named from the heavily marked veins on the undersides of its wings; the appearance of green is actually an optical illusion achieved by black scales on a yellow background. The Green-veined White rears a succession of broods through the year. One of its secrets is a successful mating plan – the female is promiscuous and the male generous, passing on a packet of protein with its sperm as well as a sprinkling of lemon-scented 'love-dust'. It is among the most widespread of our butterflies.

HOLLY BLUE, *Celastrina argiolus*

Most of our dozen species of blue butterflies inhabit wild downs and flowery banks. Only one is common in gardens and that is the Holly Blue, a pretty powder-blue butterfly with white undersides. It is double-brooded, flying in the spring and again in late summer, often along ivy-covered walls and hedges at about head height. Its plump, sluggish caterpillars feed not on leaves but on green holly berries in the spring and ivy buds in the summer. Unlike other garden butterflies, the Holly Blue rarely visits flowers; it prefers instead to sup on the sticky 'honeydew' left on leaves by feeding aphids.

MEADOW BROWN, *Maniola jurtina*

This is the most widespread of the browns, a sub-family of mostly sombre-coloured butterflies whose caterpillars feed on wild grasses. The Meadow Brown is the quintessential brown butterfly of midsummer meadows with mud-coloured wings enlivened by an eye-spot in the corner of the forewings. Time was when you could expect to find Meadow Browns in almost any patch of tall grass. Today, like so many butterflies, it has been pushed to the margins by intensive agriculture but you still find the butterfly across Britain on rough banks and downs, in disused quarries and field headlands. The Meadow Brown has been studied intensively since its wing pattern is variable and influenced by genes. Unlike most butterflies, it can fly in dull weather; perhaps brown wings help to keep a butterfly warm.

GATEKEEPER OR HEDGE BROWN, *Pyronia tithonus*

Smaller and more brightly coloured than the Meadow Brown, the Gatekeeper likes flying along thick, brambly hedges that border grass fields. Ragwort and ox-eye daisy are among its favourite flowers. It visits gardens and, being one of the last species to emerge, is the signature butterfly of late summer. It remains on the wing through August, still fresh when most other browns are becoming thin and worn.

RINGLET, *Aphantopus hyperantus*

This is our duskiest butterfly, dark chocolate-brown above but with a surprise underneath: a curved line of beautiful little rings inset with white dots. It is a quiet butterfly more tolerant of damp and shade than most species. It can even fly in light rain.

Feeding on coarse, tussocky grass such as cock's-foot and false brome, the Ringlet has adapted well to changes in the landscape such as newly planted trees or shrub-invaded grassland. It is widespread in England and is more local but increasing in lowland Scotland. Its signature ring markings vary from large and almond-shaped 'eyes' to tiny ring-less dots.

SMALL HEATH, *Coenonympha pamphilus*

Smallest of the browns, this penny-sized butterfly is easily recognised by its pale golden-brown colour. Formerly among our commonest and most widespread species on wild grassland and heath, it has decreased in many places. The caterpillar feeds on fine grasses which tend to get crowded out when regular grazing ceases or when scrub closes in. The Small Heath flies low and you often see it bobbing just above the ground or settling on the path. It has an unusually long flight season running all the way from May to September.

SPECKLED WOOD, *Pararge aegeria*

This is our woodland brown. With its pattern of light brown or yellow over dark brown the Speckled Wood is well camouflaged in the dappled light filtering through the canopy. An aggressive insect, it will defend its corner of leafy glade against all comers. Like several other browns it has increased its range and is now found in suitable places over most of lowland Britain, even visiting gardens. When not pursuing intruders, the Speckled Wood is often seen basking on a leaf or feeding on the flowers of the woodland edge, especially brambles. It has a long flight season with a succession of broods, from May to September or even later.

SMALL SKIPPER, *Thymelicus sylvestris*

Long grass is the home of this, the commonest of the skippers. Golden-brown in colour, with a dark dash-shaped 'sex-brand' on the forewing of the male, it lives up to its name with a darting, hovering, skipping flight; the Small Skipper can even fly sideways. Like several other 'golden skippers', it basks in a distinctive way with the lower wings held flat and the upper pair at an angle. Like most of our eight skippers, the Small Skipper lays its eggs on grass, usually Yorkshire fog. It is found in July over most of lowland England and Wales and is slowly edging northwards.

ESSEX SKIPPER, *Thymelicus lineola*

Most butterflies are easy to identify on the wing, but the Essex Skipper, being very similar to the Small Skipper, demands a close look. The only sure way to tell them apart is from the coloured tips of their antennae: with the Essex Skipper they are black, as though dipped in ink, while those of the Small Skipper are pale brown. The Essex Skipper was first discovered in that county but is now widespread in rough grassland across most of lowland England.

LARGE SKIPPER, *Ochlodes venata*

The Large Skipper is not, in fact, much bigger than the Small Skipper but looks quite different in its attractive mottled pattern of light and dark brown. Fresh ones can look almost golden in flight. Male Large Skippers are a delight to watch as they guard their territory, such as a patch of tall grass or a sunny hedge-bank, from their chosen perch on a leaf or grass head, darting out in a circular flight and then back again. It is a butterfly of open woodland and scrubby banks, flying in June and early July. Unlike most skippers, it visits gardens.

COMMON BLUE, *Polyommatus icarus*

Though, as its name implies, this is the commonest of the blues, the Common Blue is no longer generally common. Its stumpy caterpillar feeds on bird's-foot trefoil and that tends to define the places where you find the butterfly: sunny downs, banks, coastal cliffs and dunes, wet meadows and railway cuttings. Only the male is a clear shiny blue (often with a touch of lavender); the duller female is brown with variable orange spots and streaks of darker blue. In the south the butterfly has a long flight season, from mid-May to the end of summer.

SMALL COPPER, *Lycaena phlaeas*

This fiery little butterfly likes dry, flowery places, such as downs and heaths. You rarely see more than half a dozen together except sometimes in gardens in late summer. The Small Copper lays its eggs on sorrel but the adult has a fondness for flat, daisy-like flowers such as ragwort and fleabane. One fairly common form has iridescent blue spots on the hindwing: butterfly jewellery. It is found over most of lowland Britain and, being double-brooded, has a long flight season lasting from May through to September with a short gap in between.

PAINTED LADY, *Cynthia (or Vanessa) cardui*

Like its relative the Red Admiral, the Painted Lady is a famous long-distance migrant, found all over the world. Generally it is much less common than the Red Admiral but about once a decade the butterfly has a truly glorious season, as it did in 1996 and 2009.

Painted Ladies often share the same buddleia blossom with Peacocks and Small Tortoiseshells, their powerful fawn, black-tipped wings held stiffly to catch the sun. Its spiky caterpillar feeds on thistles. Most of the butterflies that visit Britain in early summer emerged from the chrysalis far away in the desert edges of Morocco. As befits such an epic traveller, the Painted Lady flies fast and straight but once settled on a thistle head or a buddleia it becomes completely absorbed and can be approached closely.

More Localised Butterflies

Our next group are butterflies that are less likely to visit gardens and more characteristic of wilder places, whether grassland, woodland or some other natural habitat. As a boy I thought of them as 'holiday butterflies' because that is when I used to see them, on country walks and bike rides, or by the seaside.

CLOUDED YELLOW, *Colias croceus*

This is the third and last of our regular migrants, a sunny yellow butterfly with a fast, powerful flight. In most years, Clouded Yellows are commonest along the south coast or the southerly downs of England. In good years – 1983 and 2000 were particularly good – they fly far inland and can turn up almost anywhere. In flight they look like pelting golden guineas. It is hard to get a good view of the dark-bordered wings, though, because the Clouded Yellow always rests with its wings tight shut. A proportion of the females are creamy white or pale grey. They visit us from July onwards for as long as warm weather lasts. The butterfly lays on vetches; in the past, fields of clover and lucerne grown for fodder were famous for attracting them. Recently, adult butterflies have hibernated successfully in sheltered places along the south coast to emerge in early spring.

DINGY SKIPPER, *Erynnis tages*

I always feel sorry for the Dingy Skipper which is routinely described as our dullest butterfly. It is in fact rather pretty when fresh, with its textured greyish wings and intricate pattern of dark and white spots. Unfortunately these soon fade and the wings turn a dirty grey-brown with age. The Dingy Skipper flies in the late spring. You may see it rising up at your feet and darting away, only to settle on the path a little further along. It likes warm, sheltered places with plenty of its food plant, bird's-foot trefoil.

CHALKHILL BLUE, *Polyommatus (or Lysandra) coridon*

The Chalkhill Blue is named after its habitat: the chalk and soft limestone downs of southern and eastern England. Its silvery blue wings, the colour of an English sky, have dark edges and chequered margins with a sprinkling of eye-spots underneath. The squat green caterpillar feeds only on horseshoe vetch, but the butterfly enjoys feeding on a wide variety of flowers in late summer including ground thistles, knapweed and scabious. Collectors used to love the Chalkhill Blue for its unusual variety of genetic forms, all of which they assiduously catalogued, like rare antiques.

ADONIS BLUE, *Polyommatus bellargus*

Rarer than the Chalkhill Blue, the Adonis Blue is nonetheless often common where found. It is the brightest of blues, a dazzling pure blue like a living sapphire. The Adonis shares the same habitat and food plant as its relative the Chalkhill Blue but avoids competition by laying at different times. While the Chalkhill has a single annual brood, the Adonis is double-brooded, appearing in May and again in August. It prefers southern, sun-baked hillsides where the turf is kept well cropped by rabbits or sheep and cattle.

SMALL BLUE, *Cupido minimus*

This is the smallest British butterfly, no bigger than a thumbnail, and, being dark as well as small, easily missed. The Small Blue is also the most parochial of butterflies, seldom straying far from its birthplace in a patch of kidney vetch on a warm, sheltered bank. Widespread in late May and June in the south, the Small Blue is much more local further north and one of Scotland's rarest butterflies. In Europe you often come across dozens of them feeding on seepages by a path, probing the mud with their little tongues. But in Britain they seem to prefer fresh rabbit droppings!

BROWN ARGUS, *Aricia agestis*

This is a blue that is, in fact, brown, though when freshly emerged it has an iridescence that appears silvery or blueish in flight. The Brown Argus is widespread in the southern half of Britain in rough, flowery grassland. It was once regarded as another downland butterfly, but it has also colonised quarries and other waste places by widening its choice of caterpillar food plants from rock rose to include storksbills and cranesbills. North of the Midlands the Brown Argus is replaced by its close relative, the Northern Brown Argus.

NORTHERN BROWN ARGUS, *Aricia artaxerxes*

This northern butterfly was only clearly distinguished from the Brown Argus in the twentieth century. Most have a white dot in their forewings but they are otherwise very similar to their southern cousin and also have similar habits, including the same larval food plant, rock rose. The Northern Brown Argus flies in June and July on sheltered hillsides in northern England and eastern Scotland. Those in Durham, where the white spot is faint or absent, were known as the 'Castle Eden Argus'.

MARBLED WHITE, *Melanargia galathea*

Despite its name, the Marbled White is not a true white but a brown. Like other browns its caterpillar feeds on grasses and the butterfly flies in open, grassy places in July, especially chalk and limestone hillsides. Its bright pattern, chequered like a chessboard, is unmistakable. Being toxic, this butterfly can afford to draw more attention to itself than other, more edible, browns. This is another butterfly that is gradually extending its range northwards. The Marbled White has a lazy, looping flight and often perches on a favourite flower, such as scabious or knapweed, with its wings open. Few species are quite so easy to approach and photograph.

GREEN HAIRSTREAK, *Callophrys rubi*

This is our only butterfly with truly green wings though that colour is confined to the undersides. Unobtrusive and easily overlooked, it is found in small numbers in an exceptionally wide range of habitats from chalk downs to open woods and heaths. The reason for this is the wide range of food plants available for its attractive green-and-yellow caterpillar: gorse on commons and heaths, trefoils on the downs, bilberry on moors, bramble in woods. The Green Hairstreak is a delight to watch in late spring, flashing green and brown in a hopping, jinking flight or resting with wings tight closed with its green underwear on show. Should your hand be sweaty it will readily sit there probing your skin with its little tongue.

Purple Hairstreak, *Neozephyrus quercus*

Our other four species of hairstreak are woodland or wood-edge butterflies which spend most of their adult lives high up in the woodland canopy. The Purple Hairstreak lays on the twigs of oak and is found in suitable woods throughout Britain except in northern Scotland. The small, dark, purple-shot butterfly has a jinking flight that is most easily followed with binoculars. Like other hairstreaks it is often more easily detected by looking for the small white eggs or the brown caterpillar than the adult butterfly.

White-letter Hairstreak, *Satyrium w-album*

The elusive White-letter Hairstreak is so called from the wobbly 'w' mark on the underside of its hindwing. The caterpillar feeds only on elm or wych elm in woods and hedgerows. Fortunately, despite elm disease, the butterfly is holding on reasonably well in England, especially in the Midlands. The butterfly can be watched with binoculars as it jinks or walks around a flowering elm tree. If you are lucky you might see one or two nearer the ground, sipping nectar from bramble blossom.

BROWN HAIRSTREAK, *Thecla betulae*

The Brown Hairstreak is even more elusive than the White-letter, another butterfly of the treetops – usually oak or ash – which occasionally descends to feed on flowers on hot days in August. The attractive female, which has golden-brown patches on its dusky wings, lays on blackthorn hedges. One reason for its rarity may be that such hedges are routinely flailed, resulting in the loss of most of the overwintering eggs.

DARK GREEN FRITILLARY, *Argynnis aglaja*

We have nine species of bright orange-brown butterflies speckled with black and known as fritillaries. The second-largest and overall the commonest is the Dark Green Fritillary, named after the beautiful green hindwings inset with silver 'pearls' (though the shade of green is not 'dark' but spring-fresh). The elegant, fast-flying butterfly is found in July on flower-rich grasslands across much of Britain though it is most frequent on the downs and along the coast. Like several of our fritillaries, it lays on violets – chiefly the hairy violet – but the adult butterfly prefers thistle heads.

SILVER-WASHED FRITILLARY, *Argynnis paphia*

The largest of the fritillaries is a noble butterfly with rich golden-brown wings and, beneath, pale green washed with trickles of silver. Its undulating flight, now fluttering, now gliding, is a delight to watch, especially in courtship when the male drops showers of scent and performs aerobatics which are all the more impressive against the deep green shadow of woods in midsummer. The butterfly lays on the bark of trees but the young caterpillars descend to feed on violets on the woodland floor. The Silver-washed Fritillary is more tolerant of shade than other fritillaries and that may have been its saving grace. It is still found, in July, in most of the larger woods in south-west England and more locally elsewhere. Those with long memories may recall its superabundance in that hot year, 1976. It could happen again.

WHITE ADMIRAL, *Limenitis camilla*

Sometimes sharing the same bramble patch as the Silver-washed Fritillary is this equally elegant butterfly. Its gliding flight, skimming the contours of the trees, seems effortless. The butterfly is equally striking when at rest, its austere white stripes on a dark-brown background contrasting with the paler browns and blue-greens of its undersides. Though usually seen only in ones and twos, in late June and July, the White Admiral is fairly widespread in woods across southern and eastern England. Its spiky green caterpillar feeds on honeysuckle in the half-light of deep woods.

GRAYLING, *Hipparchia semele*

The largest of the browns, the Grayling is a butterfly of warm, dry places, especially heathlands and sheltered parts of the coast. A late summer butterfly, you often spot it as it rises in front of you, soon settling down again further along the path. Its greyish hindwings form an effective camouflage on sandy paths and rocks, and the butterfly can angle its wings so that there is no shadow. It seldom visits flowers, apart from Bell Heather, but will drink from a salty puddle or even the sticky resin on a fence-post. The pale brown caterpillar feeds on bentgrasses and, unusually for butterflies, makes an underground chamber for its chrysalis.

WALL (or WALL BROWN), *Lasiommata megera*

The Wall was once a common butterfly across most of England and Wales on sunny hillsides, banks and even gardens. But it has experienced a steep and still mysterious decline and is now frequent only by the coast, on steep chalk hills and on spoil heaps in the industrial north where there is plenty of the fine-leafed fescue on which its slim green caterpillar feeds. The butterfly likes to bask on bare ground in full sun, often on footpaths. Close up it is unmistakable – and the pattern does resemble a brick wall – but in flight the Wall can be mistaken for a Comma or even a fritillary.

SILVER-STUDDED BLUE, *Plebejus argus*

The kind of dry heaths loved by the Grayling are also the home of this pretty blue, so named from the pinheads of iridescent scales on the undersides of its hindwings. Although it can be very common where found, the Silver-studded Blue is more or less restricted to heathlands and coastal headlands in the southern half of Britain where it flies in July and early August. The caterpillar feeds on gorse and other heathland plants and is looked after by ants who take the chrysalis down into their underground nest for safekeeping.

GRIZZLED SKIPPER, *Pyrgus malvae*

The smallest of the skippers and our second-smallest butterfly, the Grizzled Skipper is a chequered, moth-like butterfly whose wings become a grey blur in flight. It is a feisty little insect which engages in aerial dog-fights with rival males over its chosen patch of crumbling soil or bank. Though much declined, the Grizzled Skipper occurs locally across England and South Wales especially in partially wooded quarries and crumbling banks where its larval food plant, wild strawberry, grows.

Butterflies You Might Have to Travel a Long Way to See . . .

PURPLE EMPEROR, *Apatura iris*

Purple Emperor watching has become a popular pursuit. Many make the journey, in early July, to one of of its known localities in the East Midlands and south where you have a reasonable chance of seeing one. This is a glamourous butterfly, large, dark and strong, with a white band flickering down the wings. Only the male displays a purple iridescence; the slightly larger and more elusive female is plain dark brown and white. The butterfly spends most of its life in the treetops but it sometimes descends to feed in puddles or to bait, such as rotting banana skins – or to lay on sallow bushes. The Emperor is also attracted to fresh droppings, putrefying rabbits or even shiny car roofs – but not flowers. Its horned caterpillar feeds on sallows in the shade of oaks. Like your first sight of a whale or an eagle, you never forget a Purple Emperor.

SWALLOWTAIL, *Papilio machaon*

You never forget your first Swallowtail either, skimming majestically over the Norfolk reeds in May and June with lazy beats of its tiger-striped, elegantly tailed wings. This is our largest resident butterfly, an endemic race named *britannicus* which has darker markings than usual. Our Swallowtail is an extreme specialist, tied to the Norfolk Broads and the nearby coast by the rarity of its food plant, milk parsley, even though, in captivity, its gaily striped caterpillar will happily accept carrot tops. The butterfly often flies over water but favours open reed beds and wet, flowery meadows. The paler continental form of the Swallowtail is visiting us in increasing numbers and may soon become a settler on the south coast.

PEARL-BORDERED FRITILLARY, *Boloria euphrosyne*

This beautiful springtime fritillary, named from the row of silver spots under the hindwings, was once common in open woods throughout most of Britain. Unfortunately it has suffered a catastrophic decline and is now largely restricted to the western half of England and parts of Wales and Scotland. The reasons may be complex but the increased shading of so many woods and the decline of its caterpillar's food plant, violets, are probably the main triggers. Many of its remaining sites are nature reserves.

SMALL PEARL-BORDERED FRITILLARY, *Boloria selene*

The Small Pearl-bordered flies a few weeks later than its close relative, the Pearl-bordered, and so is usually still fresh while the latter is fading. But you need a good look at the arrangement of 'pearls' and spots on the undersides to be sure. The Small Pearl-bordered relies less on woodland than its relative, flying in damp grassland and heath wherever its larval food plant, marsh violet, occurs. Today it has become a western species in England but is more widespread in Scotland.

HIGH BROWN FRITILLARY, *Argynnis adippe*

This too was once a fairly common butterfly in woodland glades over lowland England and Wales. But today the High Brown Fritillary is rare and confined to six areas in western England and South Wales where it inhabits sheltered, open places in the vicinity of woods. The interiors of our shady woods are now too cool to support this warmth-loving butterfly and its equally warmth-loving, basking caterpillar. The latter feeds on violets among dry fronds of bracken but the fast-flying butterfly visits flowers, especially brambles.

MARSH FRITILLARY, *Euphydryas aurinia*

This is yet another fritillary in decline and one now found mainly in the western half of Britain (it is more widespread in Ireland). The Marsh Fritillary has two contrasting habitats: wet pasture and the dry chalk downs of Wiltshire and Dorset (where its name is inappropriate). Prettily marked in orange, yellow and black, this butterfly's numbers fluctuate from one year to the next depending on the activity of its parasites. It flies in June. Most colonies are small and the Marsh Fritillary is increasingly reliant on conservation management for its survival.

GLANVILLE FRITILLARY, *Melitaea cinxia*

Today this beautiful butterfly has only a toehold in Britain. It is confined to sunny coastal slopes on the Isle of Wight, especially where the clay cliffs have slumped towards the sea. This is a gem of a butterfly with a carnival of golden-yellow spots and bands on its underside, made lovelier still by its maritime setting. Its black, spiky caterpillars live in a communal nest on plantains and so are quite easy to find. The Glanville Fritillary has a relatively short flight period from mid-May to early June.

HEATH FRITILLARY, *Mellicta athalia*

The Heath Fritillary has long been rare and confined to a few open woods and sheltered valleys in Kent and the West Country. The butterfly thrives best in woods managed by regular cutting to maintain clearings full of flowers, including cow wheat and plantain on which the caterpillars feed. At one time its survival seemed uncertain but, fortunately, the Heath Fritillary has responded well to conservation management and in a few woods it can be the commonest butterfly during its flight time in late May and June, sometimes appearing in hundreds.

BLACK HAIRSTREAK, *Satyrium pruni*

The Black Hairstreak is the rarest of our five hairstreaks, being confined to woodland in the East Midlands between Oxford and Peterborough. Like its distant relative, the Brown Hairstreak, it lays its eggs on blackthorn bushes in and around woods. The butterfly flies in June and, if you are lucky, you may find it in numbers feeding on bramble or privet blossom.

SCOTCH ARGUS, *Erebia aethiops*

The Scotch Argus is not related to the Northern Brown Argus but it is confined to Scotland, apart from two colonies in the Lake District. It is an attractive deep-brown butterfly with orange bands and chequered wing margins, which frequents damp grasslands and scrub in late July and August. The males flutter tirelessly above the grass tussocks while the much more retiring females choose clumps of purple moor-grass to lay their tiny, speckled eggs.

Mountain Ringlet, *Erebia epiphron*

This small brown butterfly is our only true mountain species, occurring locally on slopes and hilltops in the Lake District and the western Grampians. It flies in June and early July, depending on the season, and only in sunshine. Its caterpillar feeds on mat-grass and, if the summer is short, it may take two years to develop. The Mountain Ringlet is one of our most elusive butterflies. It is also the one most at risk from climate change.

Large Heath, *Coenonympha tullia*

The warm brown Large Heath is one of our few wetland butterflies, occurring in peat bogs from North Wales and the Midlands northwards. It is also among the most variable of butterflies with three distinct forms that were once regarded as different races or even species. Though rather elusive, it can be quite common should you happen to find yourself in the right kind of bog on a rare sunny, windless day in June or July. The butterflies are attracted to the flowers of cross-leaved heath but the striped green caterpillar feeds on cotton-grass.

Wood White, *Leptidea sinapis*

The Wood White is a dainty butterfly with the slowest, floppiest flight of any species. It flies in May with a smaller later brood and occurs mainly in large woods or wooded cliffs in the southern half of Britain. No more than fifty colonies are known and some of those are small. The caterpillar feeds on a variety of vetches and trefoils. In Ireland a closely related species, the Cryptic Wood White, *Leptidea juvernica*, is more widespread and not confined to woodland.

DUKE OF BURGUNDY, *Hamearis lucina*

This small brown, chequered butterfly is the only member of its family, the metalmarks, in Europe. Like so many woodland butterflies it has suffered a severe decline and now occurs mainly in sheltered, scrubby limestone grassland and in open woodland in south and west England. Its nocturnal caterpillar used to feed mainly on primrose in coppiced woods but today you are more likely to find it on cowslips in the open. The Duke flies in May, and if you manage to spot as many as half a dozen you are doing well.

SILVER-SPOTTED SKIPPER, *Hesperia comma*

This stout-bodied butterfly flies close to the ground on hot August days with a fawn blur of wings. Though very local and confined to southern England it is sometimes found in good numbers on the slopes of chalk downs in full sun, especially where lumps of chalk and bare earth appear among short tussocks of fine grass. It has the ugliest of caterpillars, described by Jeremy Thomas as 'a brown-green wrinkled maggot'.

LULWORTH SKIPPER, *Thymelicus acteon*

As its name implies, this is a Dorset butterfly and one more or less confined to the coast. Here it thrives in sheltered valleys where there is plenty of its larval food plant, tor-grass. The butterfly is darker than the Small Skipper, with which it often flies, and the female has an attractive circle of golden spots on its forewings.

CHEQUERED SKIPPER, *Carterocephalus palaemon*

Extinct in England, the Chequered Skipper is still found in a limited part of Scotland, in Argyll, where it flies in small numbers in open woodland and sheltered wet hillsides in late spring. It is the prettiest of the skippers, with light brown spots over a dark brown background, and has a liking for blue flowers such as bugle and bluebell. The caterpillar feeds only on purple moor-grass.

LARGE BLUE, *Maculinea arion*

The native Large Blue died out in 1980 but a similar-looking race from Sweden has been introduced with some success. There are now around twenty-five sites where small colonies of this inky-blue butterfly have been established, all of them in south-west England. The Large Blue is wholly reliant on a species of ant that carries its caterpillar into its nest to spend the rest of its development underground. Hence, to survive, the butterfly needs not only its food plant, wild thyme, but plenty of its host ants nearby.

LARGE TORTOISESHELL, *Nymphalis polychloro*s

Although the Large Tortoiseshell was last seen in any numbers more than sixty years ago, it is possible there are small breeding colonies on the Isle of Wight and perhaps elsewhere. The butterfly, whose caterpillar feeds on elm, was always elusive even though it sometimes visited gardens. It is also possible that it may one day recolonise through natural means.

EXTINCT BUTTERFLIES

Three former British butterflies are now extinct. These are the Large Copper, *Lycaena dispar*; Mazarine Blue, *Cyaniris semi-argus*; and Black-veined White, *Aporia crataegi*. A Dutch race of the Large Copper was successfully introduced to the Fens for a while but the colony has since died out. Undoubtedly more attempts will be made in future, most likely in the Norfolk Broads.

RARE MIGRANT BUTTERFLIES

Eight rare migrants are traditionally counted as British. These are: Queen of Spain Fritillary, *Issoria lathonia*; Camberwell Beauty, *Nymphalis antiopa*; Long-tailed Blue, *Lampides boeticus*; Short-tailed or Bloxworth Blue, *Cupido* (or *Everes*) *argiades*, Bath White, *Pontia daplidice*; Pale Clouded Yellow, *Colias hyale*; Berger's Clouded Yellow, *Colias alfacariensis*; and the Monarch, *Danaus plexippus*. After its surprise showing in summer 2014 we may have to add one more species, Yellow-legged Tortoiseshell, *Nymphalis xanthomelas*. Some of these scarce migrants have occasionally laid eggs here and even produced home-bred butterflies. Thanks to butterfly hotlines, the chances of lucky sightings have greatly improved.

Notes

INTRODUCTION

1. George Monbiot, 'Back to Nature'. Commissioned online essay by the BBC at www.bbc.com/earth/bespoke/story.
2. Robert Browning (1845), 'The Lost Leader'. The poem was famously used against Harold Macmillan by one of his former Cabinet colleagues Nigel Birch during the Profumo scandal in 1963.

CHAPTER 2

1. Also the author of my favourite guide to collecting and rearing, R. L. E. Ford (1963), *Practical Entomology*. Wayside and Woodland series, Warne, London.
2. Matthew Oates (2005), 'Extreme butterfly-collecting: A biography of I. R. P. Heslop'. *British Wildlife*, 16(3), 164–71.
3. Beth Fowkes Tobin (2014), *The Duchess's Shells: Natural history collecting in the age of Cook's voyages*. Yale University Press, p. 267.
4. Reported at www.insectnet.com Forum.
5. The RCK collection was formed in 1947 and is supported by the Cockayne Trust, a charity promoting the use of the collection as a scientific resource. Details at the museum website, www.nhm. ac.uk.

CHAPTER 3

1. Facsimile edition of Thomas Moffat (1967), *The Theatre of Insects or Lesser living Creatures*. Da Capo Press, New York, p.

970. A short account of the discovery of the flattened tortoiseshell butterfly is on p. 103 of George Thomson's edition of the butterfly sections, suggesting that, assuming it to be contemporary with the manuscript, 'it is by far the oldest extant specimen of Lepidoptera'. Thomson, George C. (2012), *Insectorum sive Minorum Animalium Theatrum*. The Butterflies and Moths. Second edition. Self-published, Waterbeck, Scotland. p. 103. The MS is in the care of the British Library, Sloane 4014.

2. Petiver's directions for collectors are reproduced in Michael A. Salmon (2001), *The Aurelian Legacy*. Harley Books, Colchester, p. 57. The same source has a leaf from Buddle's butterflies on p. 59 and the original Brown Hairstreak caught or bred by Petiver on p. 58.

3. Mike Fitton and Pamela Gilbert, 'Insect Collections', in Arthur MacGregor (ed. 1994), *Sir Hans Sloane: Collector, Scientist, Antiquary, Founding Father of the British Museum*. British Museum Press, pp. 112–22. Specimen collecting of any kind seems to have begun around 1700; there is, for example, no surviving taxidermy before that date.

4. C. H. Smith (1842), 'Memoir of Dru Drury', cited in Salmon, op. cit., p. 214.

5. 'Everything is kept in true English fashion in prodigious confusion in one wretched cabinet and in boxes . . .' From a German visitor, Zacharias von Uffenbach, to Petiver's lodgings in 1710, quoted in David Allen (1976), *The Naturalist in Britain: A Social History*. Allen Lane, London, p. 38.

6. Fitton and Gilbert, op. cit., pp. 112–22.

7. Fitton and Gilbert, op. cit., p. 112.

8. Details of Ray's butterflies and their captors in C. E. Raven (2nd edn, 1950), *John Ray: Naturalist*. Cambridge University Press, pp. 407–18.

9. Vladimir Nabokov's short story, 'The Aurelian', first appeared in *Atlantic* magazine, November 1941; available online at www.theatlantic.com/magazine/archive.

10. Henry Baker 1698–1774, see Thomas Finlayson Henderson essay in *Dictionary of National Biography*, online at wikisource.org/wiki/Baker_Henry.

11. See Victoria and Albert Museum website, www.vam.ac.uk/
    content/articles/j/about_James_Leman.
12. Biography by Blanche Henrey (1986), *No Ordinary Gardener:
    Thomas Knowlton 1691–1781*. Natural History Museum,
    London.
13. Harris's account of the Cornhill fire is in Salmon, op. cit., p. 33.
14. Short biographies of Henry Leeds and other collectors in
    Salmon, op. cit.
15. E. B. Ford (1945), *Butterflies*. Collins New Naturalist, No. 1,
    p. 270.

CHAPTER 4

1. Michael A. Salmon (2001), *The Aurelian Legacy*. Harley Books,
   Colchester, pp. 68–9. A fine-art facsimile of the *Romance of
   Alexander* was published in 2014 by Quaternio Verlag, Lucerne.
2. Quoted in Christina Hardiment (2005), *Malory: The Life and
   Times of King Arthur's Chronicler*. HarperCollins, London, p. 67.
3. Letter from H. Ramsey Cox to the *Entomologist*, 11 April 1875,
   and replies. Quoted in Michael A. Salmon and Peter J. Edwards
   (2005), *The Aurelian's Fireside Companion*. Paphia Publishing,
   Lymington, pp. 31–4.
4. For example, www.britishbutterflies.co.uk/collecting.
5. Introduction by John Fowles in Kate Salway (1996), *Collector's
   Items*, Wilderness Editions, pp. 8–9.
6. Ibid.
7. Ibid.
8. Brian Boyd and Robert Michael Pyle (2000), *Nabokov's
   Butterflies*. Allen Lane, Penguin Press, London. Some of my
   notes on Nabokov and *Lolita* come from a Radio 3 programme
   on the great man broadcast on 25 April 1999. See also Kurt
   Johnson (2001), *Nabokov's Blues: The Scientific Odyssey of a
   Literary Genius*. McGraw-Hill, New York.
9. Tony Juniper (2013), *What Has Nature Ever Done for Us? Why
   money really does grow on trees*. Profile Books.
10. John Fowles, 'Lessons of Lepidoptery', The *Spectator*, 21 April
    2001, p. 40.

11. For the association with serial killers see Wikipedia.org/wiki/
The_Collector.
12. On collecting abroad, see www.theskepticalmoth.com/
collecting-permits, updated August 2014. For light relief, see
Torben B. Larsen (2004), *Hazards of Butterfly Collecting*.
Cravitz Publishing Co, Brentwood, Essex.

CHAPTER 5

1. A facsimile edition of *The Aurelian* edited by Robert Mays was
published in 1986 by Littlehampton Book Service.
2. W. S. Bristowe (1967), 'The Life of a Distinguished Woman
Naturalist, Eleanor Glanville (circa 1654–1709)', *Entomologists'
Gazette*, 18, 202–11; Bristowe (1975), 'More about Eleanor
Glanville', *Ent. Gazette*, 26, 107–17.
3. Letter quoted in R. S. Wilkinson (1966), 'Elizabeth Glanville,
an Early English Entomologist', *Entomologists' Gazette*, 17,
149–60.
4. Letter quoted in ibid.
5. Reproduced in Michael A. Salmon (2001), *The Aurelian Legacy*.
Harley Books, Colchester, p. 341.
6. Wilkinson, op. cit.
7. Ibid. For details on the Petiver collection see Mike Fitton and
Pamela Gilbert, 'Insect Collections', in Arthur MacGregor (ed.
1994), *Sir Hans Sloane, Collector, Scientist, Antiquary*, British
Museum, London, pp. 112–22.
8. Wilkinson, op. cit.
9. For Charlton's hoax butterfly, see Salmon, op. cit., p. 66 (with
illustration).
10. Richard Bradley, 'Review of Entomology' (1812) *Transactions
of the Entomological Society of London*, Volume 1, 1812. For a
biography of the Duchess, see Molly McClain (2001), *Beaufort:
The Duke and his Duchess 1657–1715*. Yale University Press,
London and New Haven.
11. Beth Fowkes Tobin (2014), *The Duchess's Shells: Natural history
collecting in the age of Cook's voyages*. Yale University Press,
London and New Haven.

12. Details of the auction are in Chapter 12, 'The Dispersal of the Collection' of Tobin, op.cit.
13. G. A. Cook (2007), 'Botanical Exchanges: Jean-Jacques Rousseau and the Duchess of Portland', in *History of European Ideas*, Pergamon Press, pp. 142–56.
14. Ibid.
15. Ibid.
16. From a large literature on women and science in the eighteenth and nineteenth centuries, e.g. Ann B. Shteir (1996), *Cultivating Women, Cultivating Science: Flora's daughters and botany in England 1760–1860*. Johns Hopkins University Press, Baltimore.
17. Sam George (2010), 'Animated Beings: Enlightenment Entomology for Girls'. *Journal for Eighteenth Century Studies*, 33 (4), 487–505.
18. Ibid.
19. For Laetitia Jermyn, see Salmon, op. cit., pp. 136–7. On Maria Catlow, see David Elliston Allen (2010), *Books and Naturalists*. Collins New Naturalist, London, 112, p. 214.
20. For Emma Hutchinson, see Salmon, op. cit., pp. 160–1.
21. Lucy Evelyn Cheesman in Joyce Duncan (2002), *Ahead of Their Time: A biographical dictionary of risk-taking women*. Greenwood Press, Westport.
22. Jane Hayter-Hames (1991), *Madam Dragonfly: The life and times of Cynthia Longfield*. Pentland Press, Edinburgh.
23. W. F. Cater (1980), *Love Among the Butterflies: The travels of a Victorian Lady*. Collins, London. Natascha Scott-Stokes (2006), *Wild and Fearless: The life of Margaret Fountaine*. Peter Owen, London.

CHAPTER 6

1. Charles Abbot, see short biography in Michael A. Salmon (2001), *The Aurelian Legacy*. Harley Books, Colchester, p. 124. There is also an *Oxford Dictionary of National Biography* article by Enid Slatter (2004).
2. Portrait and short essay on Charles Rothschild and his nature conservation work in Tim Sands (2010), *Wildlife in*

Trust: A hundred years of nature conservation. Wildlife Trusts, Newark, pp. 1–12.

3. www.ohllimited.co.uk/ashtonweb.

4. Charlotte Lane, personal communication.

5. Naomi Gryn (2004), 'Dame Miriam Rothschild'. *Jewish Quarterly*, 51, 53–8.

6. Ibid.

7. Miriam Rothschild explained her stance on homosexuality to Sue Lawley on *Desert Island Discs* on 23 April 1989. Another of her reasons was the victims' vulnerability to blackmail. The forty-minute interview can be heard online at www.bbc.co.uk/radio4/features/desert-island-discs/castaway.

8. Miriam Rothschild, personal communication.

9. Miriam Rothschild and Peter Marren (1997), *Rothschild's Reserves: Time and fragile nature*. Harley Books, Colchester, and Balaban Publishers, Rehovot. Details of 'Rothschild's Reserves' are now available online at: www.wildlifetrusts.org/rothschildsreserves.

10. Miriam Rothschild (1983), *Dear Lord Rothschild: Birds, butterflies and history*. Balaban Publishers and ISI Press, Philadelphia.

11. Sands, op. cit.

12. Miriam Rothschild (1991), *Butterfly Cooing Like a Dove*. Doubleday, London, p. 67.

13. The quotation is from Nabokov's autobiography, *Speak, Memory*, quoted in Rothschild, *Butterfly Cooing*, p. 57.

14. Rothschild, *Butterfly Cooing*, p. 67.

15. Miriam Rothschild's most accessible account of the chemical defences of butterflies and moths is 'British Aposematic Lepidoptera', introductory chapter in John Heath and A. Maitland Emmet (eds 1985), *The Moths and Butterflies of Great Britain and Ireland, Volume 2, Cossidae–Heliodinidae*. Harley Books, Colchester, pp. 9–62.

16. Details in Helmut F. van Emden and John Gurdon (2006), 'Dame Miriam Louisa Rothschild 1908–2005'. *Biographical Memoir, Fellows of the Royal Society*, 52.

17. Hannah Rothschild (2009), *The Butterfly Effect*. www.hannahrothschild.com/web/images/butterflies.

18. Rothschild, *Dear Lord Rothschild*, op. cit.
19. Ibid.
20. Ibid.
21. See Rothschild, *Dear Lord Rothschild*, op. cit., pp. 177–180.
22. Cayley-Webster (1898), *Through New Guinea and the Cannibal Countries*. Fisher Unwin, London. One to take with a pinch of malarial salt.
23. Albert Meek (1913), *A Naturalist in Cannibal Land*. Fisher Unwin, London. Rothschild said of him that 'Meek is a man who faces a danger bravely and then forgets all about it.'
24. The modest Karl Jordan has received a belated, deserved and excellent biography: Kristin Johnson (2012), *Ordering Life: Karl Jordan and the Naturalist Tradition*. Johns Hopkins University Press, Baltimore.
25. Rothschild, *Dear Lord Rothschild*, op. cit., p. 152.

CHAPTER 7

1. www.insects.org. The most detailed discussion of the word 'butterfly' I have come across is a comment piece by Anatoly Liberman (2007) on 'Wilhelm Oehl and the Butterfly', http://blog.oup.com/2007/08/butterfly.
2. The escape of a white moth from the mouth of a sleeping witch is an old tradition common to both Europe and North America (where certain large moths are known as 'witches'). It is alluded to in poetry such as 'The White Moth' by Arthur Quiller-Couch (1895).
3. A full account of dragonfly folklore across the world is Jill Lucas (2002), *Spinning Jenny and Devil's Darning Needle*. Privately published, Huddersfield. Also a summary from a British context in Peter Marren and Richard Mabey (2010), *Bugs Britannica*. Chatto & Windus, London, pp. 131–9.
4. Geoffrey Grigson (1955), *The Englishman's Flora*, Dent & Sons, London, pp. 402–4.
5. Chapter XIV, 'Of Butterflies', in facsimile of the 1658 edition

of Thomas Moffet, *The Theatre of Insects or Lesser living Creatures*. Da Capo Press, New York, pp. 957–75.

6. 'Of the Use of butterflies', in Thomas Moffet, op. cit., pp. 974–5.
7. A complete translation of the British butterfly section is in C. E. Raven (1950), *John Ray: Naturalist*. Cambridge University Press, Cambridge, pp. 407–15.
8. A good summary is A. M. Emmet, 'The Vernacular Names and Early History of British Butterflies', the introductory chapter in Emmet and Heath (1989), *The Moths and Butterflies of Great Britain and Ireland*, Volume 7, (1), *The Butterflies*. Harley Books, Colchester, pp. 7–21.
9. Michael A. Salmon (2001), *The Aurelian Legacy*. Harley Books, Colchester, pp. 110–12.
10. Peter Marren (1998), 'The English Names of Moths'. *British Wildlife*, 10(1), 29–38; Marren (2004), 'The English Names of Butterflies'. *British Wildlife*, 15(6), 401–8.
11. Richard South (1906), *The Butterflies of the British Isles*. Warne Wayside and Woodland Series, London.
12. A. Maitland Emmet (1991), *The Scientific Names of the British Lepidoptera: Their history and meaning*. Harley Books, Colchester. Emmet was the authority on the meaning of names of butterflies and moths; this book, on their Latin names, is a scholarly masterpiece.
13. From Jeremy Thomas and Richard Lewington (2010), *The Butterflies of Britain and Ireland*. British Wildlife Publishing, Oxford, p. 30.
14. Ibid., p. 36.

CHAPTER 8

1. Emmet in Emmet and Heath (1989), *Moths and Butterflies*, Volume 7(1), pp. 192–93. Emmet, A. Maitland (1989), 'The Vernacular Names and early History of British Butterflies', in ibid. Volume 7 (1) , *The Butterflies*, pp. 7–21. Among modern authors who got the facts in reverse were E. B. Ford (1945) in his famous New Naturalist volume, *Butterflies*, and T. G.

Howarth (1973) in the standard text of the day, *South's British Butterflies*.

2. Emmet, op. cit., p. 192.
3. The full text of *Cadenus and Vanessa* is available on www. luminarium.org.
4. For the *Luttrell Psalter* see the British Library's website www.bl.uk/ online gallery/sacredtexts/luttrellpsalter and Michelle Brown (2006), *The World of the Luttrell Psalter*. British Library, London.
5. 'Butterfly' in Lucia Impelluso (2003), *Nature and Its Symbols*. J. Paul Getty Museum, Los Angeles, pp. 330–2.
6. Irving F. Finkelstein (1985), 'Death, Damnation and Resurrection: Butterflies as symbols in Western art'. *Bulletin, Amateur Entomologists' Society*, 44, 123–32.
7. Ibid.
8. Ibid.
9. Brian Boyd and Robert Michael Pyle (2000), *Nabokov's Butterflies: Unpublished and uncollected writings*. Allen Lane, London, p. 676.
10. Vladimir Nabokov (1962), *Pale Fire*. Modern Classics edition, Penguin. The concluding lines of Canto 4 read:

A dark Vanessa with a crimson band / Wheels in the low sun, settles on the sand / And shows its ink-blue wingtips flecked with white. / And through the flowing shade and ebbing light / A man, unheedful of the butterfly – / Some neighbour's gardener I guess – goes by / Trundling an empty barrow up the lane.

The author's commentary continues:

One minute before his death, as we were crossing from his desmesne to mine and had begun working up between the junipers and ornamental shrubs, a Red Admirable came dizzily whirling around us like a coloured flame. Once or twice before we had already noticed the same individual, at that same time, on that same spot, where the low sun finding

an aperture in the foliage splashed the brown sand with a
last radiance while the evening's shade covered the rest of
the path. One's eyes could not follow the rapid butterfly in
the sunbeams as it flashed and vanished, and flashed again,
with an almost frightening imitation of conscious play which
now culminated in its settling upon my delighted friend's
sleeve. It took off, and we saw it next moment sporting in
an ecstasy of frivolous haste around a laurel shrub, every
now and then perching on a lacquered leaf and sliding down
its grooved middle like a boy down the banisters on his
birthday. Then the tide of the shade reached the laurels and
the magnificent, velvet-and-flame creature dissolved with it.

11. The Red Admiral was one of Miriam Rothschild's 'enigmatical'
butterflies, one apparently non-toxic but possibly containing
some chemical deterrent derived from the nettle. Miriam
Rothschild (1985) in Heath and Emmet, *Moths and Butterflies*,
Volume 2, op cit.

12. Henry Walter Bates (1864), *The Naturalist on the River
Amazons*. John Murray, London.

13. Philip Howse (2010), *Butterflies: Messages from Psyche*.
Papadakis, Winterbourne, Berks, p. 170. 'I tried to create a
tabula rasa in my mind and see the designs on the wings
without preconceptions.'

14. Ibid., p. 96 and personal communication. Amplified further in
Howse (2013), 'Lepidopteran Wing Patterns and the
Evolution of Satyric Mimicry'. *Biological Journal of the
Linnean Society*, 109, 203–14, and a forthcoming paper about
the Red Admiral.

CHAPTER 9

1. Moffet's butterfly cure and essay on the use of butterflies is in
his *Theatre of Insects or Lesser Living Creatures*, pp. 974–5,
facsimile edition (1967). Da Capo Press, New York.

2. Ibid.
3. Quoted in Charles E. Raven (2ⁿᵈ ed. 1950), *John Ray Naturalist: His life and works*. Cambridge University Press, p. 407.
4. See, for example, bees.pan-uk.org/other-pollinators. Generally butterfly tongues are too smooth for pollen to stick to. As usual there are exceptions, such as the Monarch which unwittingly carries the sticky pollen of milkweed on its tongue or legs. Certain moths can be efficient pollinators such as burnets and the Hummingbird hawkmoth. Not all bees are good pollinators either.
5. Philip Howse (2010), *Butterflies: Messages from Psyche*. Papadakis, Winterbourne, Berks. pp. 12–13.
6. Malcolm Davies and Jeyaraney Kathirithamby (1986), *Greek Insects*. Duckworth, London, pp. 99–109.
7. For example, 'Butterfly' in Lucia Impelluso (2004), *Nature and its Symbols*. Paul Getty Museum, Los Angeles, pp. 330–3.
8. Davies and Kathirithamby, op. cit., with an image on p. 105.
9. One of the treasures of the British Museum. See Richard Parkinson (2008), *The Painted Tomb-Chapel of Nebamun*. British Museum or the Museum's online website www.britishmuseum.org/explore/galleries/ancient_egypt/room_61.
10. Howse, op. cit., pp. 44–5.
11. One of the treasures of the British Library. See Janet Backhouse (1983), *Hastings Hours: A fifteenth century Book of Hours made for William, Lord Hastings, now in the British Library*. Thames & Hudson, London.
12. William Gibson (1973), *Hieronymus Bosch*. Thames & Hudson, London.
13. A. Maitland Emmet (1991), *The Scientific Names of the British Lepidoptera: Their history and meaning*. Harley Books, Colchester, p. 157.
14. Irving L. Finkelstein (1985), 'Death, Damnation and Resurrection: Butterflies as symbols in western art.' *Bulletin, Amateur Entomologists' Society*, 44, 123–32.
15. Ibid.
16. Aristotle's treatise on the soul is available online at classics.mit.edu/Aristotle/soul.

17. Moffet on moths: facsimile edition of *The Theatre of Insects*, op. cit., p. 975.
18. White butterflies flew in vast numbers over the churned landscape of the Western Front thanks to the cresses – their larval food plant – that proliferated on the open ground. Just as poppies represented sacrifice so white butterflies have come to represent peace. 'Trench butterflies', on the other hand, were bits of white toilet paper fluttering in the wind.
19. Virginia Woolf (1942), *The Death of the Moth and Other Essays*. Hogarth Press, London.
20. W. H. Hudson, *Green Mansions*, quoted in Miriam Rothschild (1991), *Butterfly Cooing like a Dove*. Doubleday, London, p. 46.
21. Paul Waring (2001), *A Guide to Moth Traps and Their Use*. Amateur Entomologists' Society, London.
22. Don Marquis, 'Archy and Mehitabel', quoted in Rothschild (1991), op. cit., p. 49.
23. 'Candles', in Rothschild (1991), op. cit., p. 46.
24. There is an outline of this pair of paintings on the website of the National Gallery of Scotland, www.nationalgalleries.org/collection.
25. Rothschild (1991), op. cit, p. 131.
26. In Philip Larkin (1988), *Collected Poems*. Faber & Faber, London.
27. www.sensationalquotes.com/Mark-Twain. Apparently it was written down in one of his notebooks.

CHAPTER 10

1. From a detailed Wikipedia article on Georg or Joris Hoefnagel.
2. Chapter on Merian by Susan Owens in David Attenborough and others (2007), *Amazing Rare Things: The art of natural history in the age of discovery*. Royal Collection Publications, London, pp. 138–75.
3. Ernest Radford, *Dictionary of National Biography*, www.oxforddnb.com. Information on his Hanoverian origins at Christies website, www.christies.com/lotfinder/lotdetails.

4. Michael A. Salmon (2001), *The Aurelian Legacy: British Butterflies and their Collectors*. Harley Books, Colchester, pp. 110–12.

5. Facsimile edition of *The Aurelian*, edited by Robert Mays (1986). Newnes Country Life Books.

6. The full title of Harris's booklet is 'The Natural System of Colours Wherein is displayed the regular and beautiful Order and Arrangement, Arising from Three Primitives, Red, Blue and Yellow. The manner in which each Colour is found, and its Composition, the Dependence they have on each other, and by their Harmonious Connection are produced the teints, or Colours, of every Object in the Creation.'

7. Howard Leathlean (2004), 'Henry Noel Humphreys', *Oxford Dictionary of National Biography*.

8. Salmon, op. cit., pp. 138–9.

9. F. W. Frohawk (1934), *The Complete Book of British Butterflies*. Ward Lock, London. It was based on the same author's massive two-volume work, *The Natural History of British Butterflies*, published by Hutchinson in 1924.

10. From Norman Riley's obituary of Frohawk in the *Entomologist*, (1947) 80, 25–7.

11. June Chatfield (1987), *F. W. Frohawk: His life and work*. Crowood Press, Ramsbury, p. 39.

12. From author's interview with Richard Lewington at his home in Oxfordshire, October 2013. For Lewington's own views on the advantages of art over photography, see Lewington (2011), 'Artwork versus Photography, Set Specimens versus Natural Posture'. *Atropos*, 43, 3–11.

13. Julian Spalding, obituary of David Measures in the *Guardian*, 12 November 2011.

14. Robert Gooden (ed., revised edn, 1981), *Beningfield's Butterflies*. Penguin Books, Harmondsworth. Obituary of Beningfield by Dennis Furnell, *Independent*, 29 May 1998.

15. Richard Tratt (2005), *Butterfly Landscapes: A celebration of British butterflies painted in natural habitat*. Langford Press, Peterborough.

CHAPTER 11

1. Mark Cocker (2013), *Birds and People*. Jonathan Cape, London, pp. 27–8.
2. Large Blue diary: ntlargeblue.wordpress.com.
3. Jeremy Thomas (2004), 'Comparative losses of British butterflies, birds and plants and the global extinction crisis'. *Science*, 1879–81.
4. Michael A. Salmon (2001), *The Aurelian Legacy*. Harley Books, Colchester, pp. 278–85. Large Coppers have come down in the world. The current price of a dozen Large Copper caterpillars from the Netherlands is about £12.50.
5. Information from www.InsectNet.com Forum.
6. www.xerces.com.
7. Josiah Clark (2009), 'A Helping Hand for the Hairstreak'. *BayNature*, baynature.org/articles.
8. www.iucnredlist.org/details/701/0.
9. Quoted in H. Mendel and S. H. Piotrowski (1986), *Butterflies of Suffolk: An atlas and history*. Suffolk Naturalists' Society, Ipswich.
10. Jeremy Thomas and Richard Lewington (2010), *The Butterflies of Britain and Ireland*. British Wildlife Publishing, Gillingham, pp. 145–7.
11. Vladimir Nabokov, 'Father's Butlerflies', in Bryan Boyd and Robert M. Pyle (eds 2001), *Nabokov's Butterflies*. Penguin.
12. L. Hugh Newman (1967), *Living with Butterflies*. John Baker, London, pp. 198–202.

CHAPTER 12

1. For the Big Butterfly Count website visit www.bigbutterflycount.org and follow the links for details.
2. In addition to the Millennium Atlas published in 2001, Butterfly Conservation has produced State of the UK's Butterflies reports in 2005 and 2011 with the next one due around 2016. They are downloadable at butterfly-conservation.org/1643/the-state-of-britains-butterflies.
3. Richard Fox et al. (2006), *The State of Britain's Butterflies in*

Britain and Ireland, downloadable at the Butterfly Conservation website or in published form by Pisces Publications, Newbury.

4. J. A. Thomas (2004), 'Comparative losses of British butterflies, birds and plants and the global extinction crisis'. *Science*, 303, 1879–81. An analysis of surveys dating back forty years indicated that 71 per cent of our butterfly species have declined over that period compared with 54 per cent of birds and 28 per cent of plants. 'No dataset approaches this detail and scale anywhere in the world,' commented Thomas.

5. 'Coast-to-coast'. *Butterfly*, 114, Autumn 2013, p. 4.

6. Martin Warren's 'breakthrough' history has not yet been written up but developed as a series of lectures to Butterfly Conservation branches. So it comes under the heading of 'personal communication'.

7. J. Heath, E. Pollard and J. A. Thomas (1984), *Atlas of Butterflies in Britain and Ireland*. Viking, Harmondsworth.

8. For the UK Butterfly Monitoring Scheme (UKBMS) visit www.ukbms.org and follow the links. The scheme is operated as a partnership between the Centre for Ecology and Hydrology (CEH), Butterfly Conservation and the Joint Nature Conservation Committee (JNCC).

9. Although Thomas's PhD thesis is unpublished and the report he based on it confidential, a good summary of his findings can be found in the Black Hairstreak section of Jeremy Thomas and Richard Lewington (2010), *The Butterflies of Britain and Ireland*. British Wildlife Publishing, Gillingham.

10. There is a large and growing literature on the conservation of the Marsh Fritillary. Keith Porter wrote up his observations on the basking behaviour of its caterpillars in *Oikos* (1997). A more accessible account of this problematic butterfly can be found in Thomas and Lewington, op. cit., and an account of the 'lessons' learned in Nigel Bourne et al. (2013), in *British Wildlife*.

11. There is already a vast and growing literature on metapopulations. Hanski's original paper on the Glanville Fritillary is Hanski, I (1994), A Practical model of

metapopulation dynamics. *J. Animal Ecology*, 63, 151–62. *See Also* Hanski, I (2003) Biology of extinctions in butterfly metapopulations. In: *Butterflies – ecology and evolution taking flight* (ed. C.L. Bloggs, W. B. Watt & P. R. Ehrlich), pp. 577–602. University of Chicago Press, Chicago.

12. See Brereton et al. (2012), in *British Wildlife*. In a wider context Butterfly Conservation has an Action Plan for the Pearl-bordered Fritillary at butterfly-conservation.org/files/pearl-bordered-fritillary-action-plan.doc.

13. Butterfly Conservation's ten-point plan for saving our butterflies is in Fox et al. (2006), pp. 104–106

14. Robert Frost, 'Blue Butterfly Day', in *Complete Poems of Robert Frost* (1949). Henry Holt, New York.

15. See Butterfly Conservation website and follow the link 'Why butterflies matter'.

CHAPTER 13

1. Jeremy Thomas and Richard Lewington (2010), *The Butterflies of Britain and Ireland*. British Wildlife Publishing, Gillingham, p. 77.

2. Easily the most entertaining butterfly website is Matthew Oates's, mainly on the Purple Emperor, 'The Purple Empire', created for 'people of the purple persuasion'. See apaturairis.blogspot.com. There is a blog by Martin Warren, Richard Fox and others on the Butterfly Conservation website, link: News and Blog.

3. As reported by Richard Fox and Nick Bowles in 'Wildlife Reports' (2014). *British Wildlife*, 26 (2), pp. 124-27.

4. Miriam Rothschild (1991), 'Silk', in *Butterfly Cooing like a Dove*. Doubleday, London, p. 67.

Bibliography

Allan, P. B. M. (2ⁿᵈ edn 1947), *A Moth Hunter's Gossip*. Watkins & Doncaster, London.

Allen, David Elliston (1976), *The Naturalist in Britain: A social history*. Allen Lane, London.

Allen, David Elliston (2010), *Books and Naturalists*. Collins New Naturalist, London.

Asher, Jim et al. (2001), *The Millennium Atlas of Butterflies in Britain and Ireland*. Oxford University Press, Oxford.

Attenborough, David et al. (2007), *Amazing Rare Things: The art of natural history in the age of discovery*. Royal Collection Publications, London.

Bourne, Nigel et al. (2013), 'Conserving the Marsh Fritillary across the UK: lessons for landscape-scale conservation'. *British Wildlife* 24 (6), 408–17.

Boyd, Brian and Pyle, Robert Michael (eds. 2000), *Nabokov's Butterflies: Unpublished and uncollected writings*. Allen Lane, London.

Brereton, Tom, Pilkington, Gary and Roy, David (2012), 'Conserving Violet-feeding Fritillary Butterflies at Marshland Nature Reserve'. *British Wildlife* 24 (1), 1–8.

Bristowe, W. S. (1967), 'The Life of a Distinguished Woman Naturalist, Eleanor Glanville (circa 1654–1709)'. *Entomologists Gazette*, 18, 202–11.

Butterflies Under Threat Team [BUTT] (1986), *The Management of Chalk Downland for Butterflies*. Focus on Nature Conservation Series, No. 17. Nature Conservancy Council, Peterborough.

Butterfly Conservation (2010), *The 2010 Target and Beyond for Lepidoptera*. 6th International Symposium. Butterfly Conservation.

Butterfly Conservation (2011), *The State of the UK's Butterflies 2011*. Online report, butterfly-conservation.org/files/soukb2011.

Chatfield, June (1987), *F. W. Frohawk: His life and work*. The Crowood Press, Ramsbury.

Conniff, Richard (2010), *The Species Seekers: Heroes, fools and the mad pursuit of life on earth*. Norton, London.

Curtis, Robin (2014), 'The Glanville Fritillary: A disappearing gem?' *British Wildlife*, 25 (6), 405–12.

Davies, Malcolm and Kathirithamby, Jeyaraney (1986), *Greek Insects*. Duckworth, London.

Dunbar, David (2010), *British Butterflies: A history in books*. The British Library, London.

Emmet, A. M. (1989), 'The vernacular Names and Early History of British Butterflies. In Heath, John and Emmet, A. Maitland (eds) (1989), *The Moths and Butterflies of Great Britain and Ireland*. Volume 7, *Hesperiidae–Nymphalidae*. *The Butterflies*. Harley Books, Colchester, pp. 7–21.

Emmet, A. Maitland (1991), *The Scientific Names of the British Lepidoptera: Their history and meaning*. Harley Books, Colchester.

Finkelstein, Irving F. (1985), 'Death, Damnation and Resurrection: Butterflies as symbols in Western art'. Bulletin, *Amateur Entomologists' Society*, 44, 123–132.

Fitton, Mike and Gilbert, Pamela (1994), 'Insect Collections'. In MacGregor, Arthur, *Sir Hans Sloane*. British Museum Press, London, pp. 112–22.

Ford, R. L. E. (1952), *The Observer's Book of Larger Moths*. Warne, London.

Fox, Richard et al. (2006), *The State of Butterflies in Britain and Ireland*. Pisces Publications for Butterfly Conservation.

Frohawk, F. W. (1934), *The Complete Book of British Butterflies*. Ward, Lock and Co., London.

Hanski, I.; Kussaeri, M. and Nieninen, M. (1994), 'Metapopulation Structure and Migration in the Butterfly *Melitaea cinxia*. *Ecology*, 75, 747–62.

Harmer, Alec (2013), 'Like Father, Like Son: the "Lost" Entomological Paintings of John Harris (1767–1832) and the Remarkable Harris legacy'. *Antenna, Bulletin of the Royal Entomological Society*, 37 (1), 4–19.

Heath, J.; Pollard, E. and Thomas, J. A. (1984), *Atlas of Butterflies in Britain and Ireland*. Viking, Harmondsworth.

Heslop, I. R. P.; Hyde, G. E. and Stockley, R. E. (1964), *Notes and Views on the Purple Emperor*. Southern Publishing Company, Brighton.

Hill, Les; Randle, Zoe; Fox, Richard and Parsons, Mark (2010), *Provisional Atlas of the UK's Larger Moths*. Butterfly Conservation, Wareham, Dorset.

Howse, Philip (2010), *Butterflies: Messages from Psyche*. Papadakis, Winterbourne, Berks.

Howse, Philip (2014), *Seeing Butterflies: New perspectives on colour, patterns and mimicry*. Papadakis, Winterbourne, Berks.

Impelluso, Lucia (2004), *Nature and its Symbols*. J. Paul Getty Museum, Los Angeles.

Johnson, Kristin (2012), *Ordering Life: Karl Jordan and the naturalist tradition*. Johns Hopkins University, Baltimore.

MacGregor, Arthur (ed. 1994), *Sir Hans Sloane: Collector, scientist, antiquary, founding father of the British Museum*. British Museum Press, London.

Marren, Peter (1998a), 'A Short History of Butterfly-collecting in Britain'. *British Wildlife*, 9, 362–70.

Marren, Peter (1998b), 'The English Names of Moths'. *British Wildlife*, 10, 29–38.

Marren, Peter (2004), 'The English Names of Butterflies'. *British Wildlife*, 15, 401–8.

Marren, Peter and Mabey, Richard (2010), *Bugs Britannica*. Chatto & Windus, London.

Matthews, Patrick (ed. 1957), *The Pursuit of Moths and Butterflies*. Chatto & Windus, London.

Mays, Robert (ed. 1986), *The Aurelian or The Natural History of English Insects, Namely Moths and Butterflies. Together with the Plants on which they Feed* by Moses Harris. Facsimile edition. Country Life Books, London.

Mendel, H. and Piotrowski, S. H. (1986), *Butterflies of Suffolk: An atlas and history*. Suffolk Naturalists' Society, Ipswich.

Newman, L. Hugh (1967), *Living with Butterflies*. John Baker, London.

Oates, Matthew (2005), 'Extreme Butterfly-collecting: A biography of I. R. P. Heslop'. *British Wildlife* 16 (3) 164–71.

Porter, K. (1997), 'Basking Behaviour in Larvae of the Butterfly, *Euphydryas aurinia*'. *Oikos*, 308–12.

Raven, C. E. (2nd edn 1950), *John Ray, Naturalist: His life and works*. Cambridge University Press, Cambridge.

Rothschild, Miriam (1983), *Dear Lord Rothschild: Birds, butterflies and history*. Balaban Publishers, Rehovot, and ISI Press, Philadelphia.

Rothschild, Miriam (1985), 'British Aposematic Lepidoptera'. In Heath, John and Emmet, A. Maitland (eds), *The Moths and*

Butterflies of Great Britain and Ireland. Volume 2, *Cossidae– Heliodinidae*, pp. 9–62.

Rothschild, Miriam (1991), *Butterfly Cooing Like a Dove*. Doubleday, London.

Rothschild, Miriam and Farrell, Clive (1983), *The Butterfly Gardener*. Michael Joseph, London.

Rothschild, Miriam and Marren, Peter (1997), *Rothschild's Reserves: Time and fragile nature*. Balaban Publishers, Rehovot, and Harley Books, Colchester.

Salmon, Michael, A. (2000), *The Aurelian Legacy: British butterflies and their collectors*. Harley Books, Colchester.

Salmon, Michael A. and Edwards, Peter J. (2005), *The Aurelian's Fireside Companion: An entomological anthology*. Paphia Publishing, Lymington.

Spalding, Julian (2011), Obituary of David Measures. *Guardian*, 12 August 2011.

Sterling, Phil; Parsons, Mark and Lewington, Richard (2012), *Field Guide to the Micro Moths of Great Britain and Ireland*. British Wildlife Publishing, Gillingham.

Stokoe, W. J. (1938), *The Observer's Book of Butterflies*. Warne, London.

Thomas, J. A. (1980), 'Why Did the Large Blue Become Extinct in Britain?' *Oryx*, 15, 243–7.

Thomas, J. A. (1999), 'The Large Blue Butterfly – A Decade of Progress'. *British Wildlife*, 11, 22–7.

Thomas, J. A. et al. (2004), 'Comparative Losses of British Butterflies, Birds and Plants and the Global Extinction Crisis'. *Science*, 1879–81.

Thomas, Jeremy and Lewington, Richard (2010), *The Butterflies of Britain and Ireland*. British Wildlife Publishing, Gillingham.

Thompson, Ken (2006), *No Nettles Required: The reassuring truth about wildlife gardening*. Eden Project Books and Transworld Publishers, London.

Thomson, George (2nd edn 2012), *The Butterflies and Moths*. From *Insectorum sive Minimorum Animalium Theatrum* by Thomas Moffet. Privately published.

Tobin, Beth Fowkes (2014), *The Duchess's Shells: Natural history collecting in the age of Cook's voyages*. Yale University Press, London.

Tolman, Tom and Lewington, Richard (2008), *Collins Butterfly Guide: The most complete guide to the butterflies of Britain and Europe*. Collins, London.

Tratt, Richard (2005), *Butterfly Landscapes: A celebration of British butterflies painted in natural habitat*. Langford Press, Peterborough.

Van Emden, Helmut F. and Gurdon, J. (2006), 'Dame Miriam Rothschild 1908–2005'. *Biographies of Members of the Royal Society*, 52, 315–30.

Vane-Wright, R. I. and Hughes, H. W. D. (2005), *The Seymer Legacy: Henry Seymer and Henry Seymer Jnr of Dorset and their entomological paintings*. Forrest Text, Cardigan.

Waring, Paul; Townsend, Martin and Lewington, Richard (2003), *Field Guide to the Moths of Great Britain and Ireland*. British Wildlife Publishing.

Warren, M. S. (1987), 'The Ecology and Conservation of the Heath Fritillary, *Mellicta athalia*'. *Journal of Applied Ecology*, 24, 467–513.

Werness, Hope B. (2006), *The Continuum Encyclopaedia of Animal Symbolism in Art*. Continuum Publishing Group, London and New York.

Acknowledgements

This will be a short thank-you list because this book has been mainly the product of one person thinking about butterflies. It would be too coy to acknowledge my butterfly muses; they are not receptive to compliments of that kind, although the Purple Emperor I bred from an overwintering larva, whose pristine beauty I rewarded by offering it a drop of sweet sherry on my finger-tip, seemed to take it in the right spirit. Even so, I feel I must put in a good word for the Red Admiral. It has unwittingly inspired writers and artists down the ages and it succeeded in inspiring me too. I dreamed up this book while watching a fine fresh female Admiral sucking the juices from a blossom of ivy on one of the last warm days of the year. God bless you, *Vanessa atalanta*.

The idea was also inspired in part by conversations with the late Miriam Rothschild whose words and sayings live on after her death in 2005 as reverberating echoes whenever the colours and patterns of butterflies are discussed. I have learned much from others equally mad about the butterfly, most notably from Matthew Oates, Jeremy Thomas and Martin Warren. Other friends have helped me, wittingly or otherwise, with intriguing ideas or snippets of information, be it last week or yesteryear, notably David Elliston Allen, John F. Burton, Tim Bernard,

Andrew Branson, David Dunbar, Richard Fox, Bob Gibbons, Alec Harmer, Basil Harley, Jack Harrison, Jeremy Mynott, Paul Raven, Michael Salmon, Matt Shardlow, Roger Smith, Brett Westwood, David Wislon, David Withrington and Mark Young.

Miriam's daughter Charlotte Lane helped me with details about the Rothschilds and kindly read and commented on what I had written about them. I am grateful to Malcolm Scoble, Julie Harvey, David Carter and Mark Parsons for showing me some of the entomological treasures of the Natural History Museum, and to George McGavin and Stella Brecknell for performing the same service at the University Museum, Oxford. I thank Philip Howse for sharing some of his insights into the mimicry and deception hidden in butterfly wings and Richard Lewington for giving up a day to showing me his studio and answering all my questions so patiently. I also thank the moderators of Insectnet.com Forum, a rich source of information about collectors and collecting the world over.

In another sense I would like to thank those friends who set me off on the trail. Mike McCarthy has a rare talent, honed over decades of journalistic activity, for asking the right question, the obvious question, and the one which is, all too often, the overlooked question. Any conversation with Richard Mabey, with whom I collaborated on *Bugs Britannica*, sets the mind racing. Certain other ideas came about through ritual, ruminative evening drinks over the kitchen table with my country neighbour John Norton who has never quite managed to outgrow his sense of wonder at the might and majesty of nature.

Turning to the production of *Rainbow Dust*, I am grateful first to my editor, Rosemary Davidson, who piloted this boat from the beginning to the end of its long voyage, and corrected my various errors in navigation. I also thank Penny Hoare for reading the whole draft and for her invaluable comments. I thank my favourite wildlife artist, Carrie Akroyd, for designing a beautiful jacket, and the book's designer, Julia Connolly, for her neat, clean pages. Mary Chamberlain was the careful and thoughtful copy editor one dreams about, and Anthony Hippisley an equally eagle-eyed proof reader. Within Random House, I also thank Kate Bland, Sara Holloway, Mikaela Pedlow, Natalie Wall, Rowena Skelton-Wallace, Simon Rhodes and Mari Yamakazi. Emily Beech at the Natural History Museum assisted us with tracking down and scanning images taken from ancient texts.

Right from the start I intended to dedicate *Rainbow Dust* to my dear friends Emma and Claire Garnett. If our outings in search of butterfly eggs and caterpillars, amid other rural escapades, had anything at all to do with their subsequent, highly promising scientific careers I would be very proud.

Index

Royal Veterinary College 98
Royal William 133
rubbish dumps 181, 248, 253
Rudolf II, Holy Roman Emperor 176–7
Ruskin, John 108, 233
Russia 68, 69–70, 149–50
  Russian Revolution (1917) 68, 69

S
'Sad brown' 33
sallow bushes 254
Salway, Kate 66
San Francisco, United States 77, 205–7
Sands, Tim 113
Satyrs 137
scabious 249
Scandinavia 76
Scarce Tortoiseshell 234
scent 114–15, 126, 235, 251
schizophrenia 105, 106–7
  Schizophrenia Research Fund 106–7
Schrank, Franz 139, 160
Scotch Argus 128, 130, 135, 138, 257–8
Scotland 90, 103, 144, 217, 244, 248, 249, 250, 255, 257, 258, 260
Scottish Chequered Skipper 103
scrubland 196, 206, 221, 245, 257, 259
Seghers, Daniel 148
Semele 138
semen 157
Sense of Wonder, The (Carson) v
setting butterflies 21–2, 29, 44, 58, 65–6, 175
Shakespeare, William 61, 159
  Macbeth 201
  Midsummer Night's Dream, A 168
  Richard III 159
sheep 72, 117, 217, 218, 248
Sherard, William 54
Shooter's Hill, London 26
Short-tailed Blue 261
Sierra Nevada, Spain 207–8
Silent Spring (Carson) 219
silk moths 27, 119
silkworms 53
silver lupin 205–6
Silver-spotted Skipper 135, 214, 223, 259

Silver-studded Blue 17, 134, 141, 209, 253
Silver-washed Fritillary 134, 197, 235, 251–2
Simcox, David 196
Singapore 208
Sites of Special Scientific Interest (SSSI) 215
Sizewell nuclear power station, Suffolk 209
Skidmore, David 39
skippers 10, 11, 29, 31, 44, 133, 139, 141
  Chequered Skipper 101–3, 135
  Dingy Skipper 133, 139, 247
  Essex Skipper 135, 244
  Grizzled Skipper 208, 253–4
  Large Skipper 133, 139, 244–5
  Lulworth Skipper 259–60
  Pearl Skipper 135
  Silver-spotted Skipper 135, 214, 223, 259
  Small Skipper 133, 139, 244
  Spotted Skipper 135
skylarks 129
slides 44
Sloane, Hans 46–8, 53, 79, 80, 84, 86, 178
Small Blue 13–14, 134, 141, 217, 248
Small Copper 132–3, 245
Small Heath 132, 243
Small Pearl-bordered Fritillary 226, 255
Small Skipper 133, 139, 244
Small Tortoiseshell 13, 42, 131, 159–60, 174, 192, 232, 238, 246
Small White 207, 232, 238
Smith, Roger 198
Society for the Promotion of Nature Reserves 110
Society of Apothecaries 51
Society of Arts 53
Society of Aurelians 51–5, 81, 102, 133–4, 182
Solander, Daniel 90
Somerset, England 80–1, 83, 86, 196, 197–200
Somerset, Mary Capel 86
sorrel 245
Sotheran's bookshop, Piccadilly 112